Institute of Mathematical Statistics

LECTURE NOTES–MONOGRAPH SERIES

Volume 39

R.R. Bahadur's Lectures on the Theory of Estimation

Stephen M. Stigler, Wing Hung Wong and Daming Xu, Editors

Institute of Mathematical Statistics
Beachwood, Ohio

Institute of Mathematical Statistics
Lecture Notes-Monograph Series

Series Editor:
Joel Greenhouse

The production of the *Institute of Mathematical Statistics
Lecture Notes-Monograph Series* is managed by the
IMS Societal Office: Julia A. Norton, Treasurer and
Elyse Gustafson, Executive Director.

Library of Congress Control Number: 2002104854

International Standard Book Number 0-940600-53-6

Editors' Preface

In the Winter Quarter of the academic year 1984-1985, Raj Bahadur gave a series of lectures on estimation theory at the University of Chicago. The role of statistical theory in Chicago's graduate curriculum has always varied according to faculty interests, but the hard and detailed examination of the classical theory of estimation was in those years Raj's province, to be treated in a special topics course when his time and the students' interests so dictated. Winter 1985 was one of those times. In ten weeks, Raj covered what most mathematical statisticians would agree should be standard topics in a course on parametric estimation: Bayes estimates, unbiased estimation, Fisher information, Cramér-Rao bounds, and the theory of maximum likelihood estimation. As a seasoned teacher, Raj knew that mathematical prerequisites could not be taken entirely for granted, that even students who had fulfilled them would benefit from a refresher, and accordingly he began with a review of the geometry of L^2 function spaces.

Two of us were in that classroom, WHW who was then a junior member of the statistics faculty and DX who was then an advanced graduate student. We both had previously studied parametric estimation, but never from Raj's distinctive perspective. What started as a visit motivated by curiosity (just how would one of the architects of the modern theory explain what were by then its standard elements?) soon became a compelling, not-to-be-missed pilgrimage, three times a week. Raj's approach was elegant and turned what we thought were shop-worn topics into polished gems; these gems were not only attractive on the surface but translucent, and with his guidance and insight we could look deeply within them as well, gaining new understanding with every lecture. Topics we thought we understood well, such as locally unbiased estimates and likelihood ratios in Lecture 11, and the asymptotic optimality of maximum likelihood estimators in Lectures 28-30, were given new life and much more general understanding as we came to better understand the principles and geometry that underlay them. The two of us (WHW and DX) took detailed notes, reviewing them after the lectures to recover any gaps and smooth the presentation to a better approximation of what we had heard. Some time after the course, Raj was pleased to receive from us an edited, hand-copied version of his lectures.

In these lectures, Raj Bahadur strived towards, and in most cases succeeded in deriving the most general results using the simplest arguments. After stating a result in class, he would usually begin its derivation by saying "it is really very simple…", and indeed his argument would often appear to be quite elementary and simple. However, upon further study, we would find that his arguments were far from elementary — they appear to be so only because they are so carefully crafted and follow such impeccable logic. Isaac Newton's *Principia* abounds with "simple" demonstrations based upon a single well-designed diagram, demonstrations that in another's hands might have been given through pages of dense and intricate mathematics, the result proved correctly but without insight. Others, even the great astrophysicist S. Chandrasekhar (1995) would marvel at how Newton saw what could be so simply yet generally accomplished. So it is with some of these lectures. Raj focused on the essential and core aspect of problems, often leading him to arguments useful not only for the solutions of the immediate problems, but also for the solutions of very large classes of related problems. This is illustrated by his treatment of Fisher's bound for the asymptotic variance of estimators (Lectures 29-30). Here he used the Neyman-Pearson lemma and the asymptotic normal distribution under local alternatives to provide a remarkably elegant proof of the celebrated result (which he attributed to LeCam) that the set of parameter points at which the Fisher bound fails must be of measure zero. Although Raj developed this theory under the assumption of asymptotic normality, his approach is in fact applicable to a much wider class of estimators (Wong, 1992; Zheng and Fang, 1994).

In this course, Raj Bahadur did not use any particular text and clearly did not rely on any book in his preparation for lectures, although he did encourage students to consult other sources to broaden their perspective. Among the texts that he mentioned are Cramér (1946), Pitman (1979), Ibragimov and Hasminskii (1981), and Lehmann (1983). Raj's own papers were useful, particularly his short and elegant treatment of Fisher's bound (Bahadur, 1964, which is (23) in the appended bibliography of Raj's works). Raj did lecture from notes, but they were no more than a sketchy outline of intentions, and none of them survive. And so these lecture notes are exactly that, the notes of students.

A couple of years after these notes were recorded Raj gave the course again, but we know of no record of any changes in the material, and he may well have used these very notes in preparing for those lectures. While Raj at one time considered using these notes as a basis for a monographic treatment, that was not to be. As a consequence they reflect very much the pace and flavor of the occasion the course was given. Some topics occupied several lectures; others were shorter. Homework problems were stated at the time he thought them most appropriate, not at the end of a topic or chapter. Notation would shift occasionally, and the one change that we have made in this published version is to attempt to make the notation consistent.

Unfortunately, many memorable aspects of Raj's teaching style cannot be conveyed in these notes. He had a great sense of humor and was able to amuse the class with a good number of unexpected jokes that were not recorded. Raj also possessed a degree of humility that is rare among scholars of his stature. His showed no outward signs of self-importance and shared his time and insight without reservation. After class his door was always open for those who needed help and advice, except, of course, immediately after lunch hour, when he was needed in the Billiard Room of the Quadrangle Club.

For many years these notes were circulated as xeroxed copies of the handwritten originals. The repeated requests for copies we received over many years led us to prepare them for publication. These lectures are now superbly typed in Tex by Loren Spice, himself an expert mathematician, and in the process he helped clarify the exposition in countless small ways. They have been proofread by George Tseng, Donald Truax, and the editors. We are grateful for the support of the University of Chicago's Department of Statistics and its Chairman Michael Stein, and for the assistance of Mitzi Nakatsuka in their final preparation. These lecture notes are presented here with the permission of Thelma Bahadur, in the hope of helping the reader to appreciate the great beauty and utility of the core results in estimation theory, as taught by a great scholar and master teacher Raj Bahadur.

Stephen M. Stigler (University of Chicago)
Wing Hung Wong (Harvard University)
Daming Xu (University of Oregon)

Table of Contents

Lectures on Estimation Theory

R. R. Bahadur
(Notes taken by Wing Hung Wong and Daming Xu
and typed by Loren Spice)

(Winter quarter of academic year 1984–1985)

Chapter 1

Note on the notation: Throughout, Professor Bahadur used the symbols $\varphi(s)$, $\varphi_1(s)$, $\varphi_2(s)$, ... to denote functions of the sample that are generally of little importance in the discussion of the likelihood. These functions often arise in his derivations without prior definition.

Lecture 1

Review of L^2 geometry

Let $(S, \mathcal{A}, P)$ be a probability space. We call two functions f_1 and f_2 on S EQUIVALENT if and only if $P(f_1 = f_2) = 1$, and set

$$V = L^2(S, \mathcal{A}, P) := \left\{ f : f \text{ is measurable and } E(f^2) = \int_S f(s)^2 dP(s) < \infty \right\},$$

where we have identified equivalent functions. We may abbreviate $L^2(S, \mathcal{A}, P)$ to $L^2(P)$ or, if the probability space is understood, to just L^2. For $f, g \in V$, we define $||f|| = +\sqrt{E(f^2)}$ and $(f, g) = E(f \cdot g)$, so that $||f||^2 = (f, f)$. Throughout this list f and g denote arbitrary (collections of equivalent) functions in V.

1. V is a real vector space.

2. $(\cdot, \cdot)$ is an inner product on V – i.e., a bilinear, symmetric and positive definite function.

3. CAUCHY-SCHWARZ INEQUALITY:

$$|(f, g)| \leq ||f|| \cdot ||g||,$$

   with equality if and only if f and g are linearly dependent.

   Proof. Let x and y be real; then, by expanding $||\cdot||$ in terms of $(\cdot, \cdot)$, we find that

   $$0 \leq ||xf + yg||^2 = x^2 ||f||^2 + 2xy(f, g) + y^2 ||g||^2,$$

   from which the result follows immediately on letting $x = ||g||$ and $y = ||f||$. $\square$

4. TRIANGLE INEQUALITY:

$$||f + g|| \leq ||f|| + ||g||.$$

Proof.

$$||f + g||^2 = |||f||^2 + 2(f, g) + ||g||^2| \leq ||f||^2 + 2||f|| \, ||g|| + ||g||^2,$$

again by expanding $||\cdot||$ in terms of $(\cdot, \cdot)$ and using the Cauchy-Schwarz inequality. □

5. PARALLELOGRAM LAW:

$$||f + g||^2 + ||f - g||^2 = 2(||f||^2 + ||g||^2).$$

Proof. Direct computation, as above. □

6. $||\cdot||$ is a continuous function on V, and $(\cdot, \cdot)$ is a continuous function on $V \times V$.

Proof. Suppose $f_n \xrightarrow{L^2} f$; then

$$(||f_n|| \leq ||f|| + ||f_n - f|| \to ||f||) \Rightarrow (\overline{\lim} \, ||f_n|| \leq ||f||)$$

and

$$(||f|| \leq ||f_n|| + ||f_n - f||) \Rightarrow (\underline{\lim} \, ||f_n|| \geq ||f||).$$

From these two statements it follows that $\lim ||f_n|| = ||f||$. □

7. V is a complete metric space under $||\cdot||$ – i.e., if $\{g_n\}$ is a sequence in V and $||g_n - g_m|| \to 0$ as $n, m \to \infty$, then $\exists \gamma \in V$ such that $||g_n - \gamma|| \to 0$.

Proof. The proof proceeds in four parts.

1. $\{g_n\}$ is a Cauchy sequence in probability:

$$P(|g_m - g_n| > \varepsilon) = P(|g_m - g_n|^2 > \varepsilon^2) \leq \frac{1}{\varepsilon^2} E(||g_m - g_n||^2) = \frac{1}{\varepsilon^2}||g_m - g_n||^2.$$

2. Hence there exists a subsequence $\{g_{n_k}\}$ converging a.e.(P) to, say, g.

3. $g \in V$.

   Proof.

   $$E(|g|^2) = \int \left(\lim_{k \to \infty} g_{n_k}^2 \right) dP \leq \underline{\lim} \int g_{n_k}^2 \, dP$$

   by Fatou's lemma; but $\{\int g_{n_k}^2 \, dP = ||g_{n_k}||^2\}$ is a bounded sequence, since $\{||g_n||\}$ is Cauchy. □

3

4. $||g_n - g|| \to 0$.

Proof. For any $\varepsilon > 0$, choose $k = k(\varepsilon)$ so that $||g_m - g_n|| < \varepsilon$ whenever $m, n \geq k(\varepsilon)$. Then

$$\int |g_n - g|^2 dP = \int \left(\lim_{k \to \infty} |g_n - g_{n_k}|^2 \right) dP$$

$$\overset{\text{Fatou}}{\leq} \lim_{k \to \infty} \int |g_n - g_{n_k}|^2 dP = \lim_{k \to \infty} ||g_n - g_{n_k}||^2 < \varepsilon,$$

provided that $n > k(\varepsilon)$. $\qquad\square$

Let W be a subset of V. If W is closed under addition and scalar multiplication, then it is called a LINEAR MANIFOLD in V. If, furthermore, W is topologically closed, then it is called a SUBSPACE of V. Note that a finite-dimensional linear manifold must be topologically closed (hence a subspace).

If C is any collection of vectors in V, then let C_1 be the collection of all finite linear combinations of vectors in C and C_2 be the closure of C_1. Then C_2 is the smallest subspace of V containing C, and is called the subspace SPANNED by C. C_1 is called the linear manifold spanned by C.

Let W be a fixed subspace of V, and f a fixed vector in V. We say that the vector $g \in W$ is an ORTHOGONAL PROJECTION of f to W if and only if

$$||f - g|| = \inf_{h \in W} ||f - h||.$$

8. There exists a unique orthogonal projection g of f to W.

Proof. Let $\ell = \inf_{h \in W} ||f - h||$, and let $\{g_n\}$ be a sequence in W such that $||f - g_n|| \to \ell$; then we have

$$\left|\left| \frac{g_m - g_n}{2} \right|\right|^2 + \underbrace{\left|\left| \frac{g_m + g_n}{2} - f \right|\right|^2}_{\geq \ell} = \frac{1}{2} \underbrace{||g_m - f||^2}_{\text{converges to } \ell} + \frac{1}{2} \underbrace{||g_n - f||^2}_{\text{converges to } \ell},$$

from which we see that $||g_m - g_n||^2 \to 0$ as $m, n \to \infty$. Thus $\{g_n\}$ is a Cauchy sequence; but this means that there is some g such that $g_n \to g$. Since W is a subspace of V, it is closed; so, since each $g_n \in W$, so too is $g \in W$. $\qquad\square$

Lecture 2

Definition. For two vectors $f_1, f_2 \in V$, we say that f_1 is ORTHOGONAL to f_2, and write $f_1 \perp f_2$, if and only if $(f_1, f_2) = 0$.

Throughout, we fix a subspace W of V and vectors $f, f_1, f_2 \in V$.

9. PYTHAGOREAN THEOREM (and its converse):
$$f_1 \perp f_2 \Leftrightarrow ||f_1 + f_2||^2 = ||f_1||^2 + ||f_2||^2.$$

10. a. Given the above definition of orthogonality, there are two natural notions of orthogonal projection:

(*) $\gamma \in W$ is an orthogonal projection of f on W if and only if
$$||f - \gamma|| = \inf_{g \in W} ||f - g||.$$

(**) $\gamma \in W$ is an orthogonal projection of f on W if and only if
$$(f - \gamma) \perp g \ \forall g \in W.$$

These two definitions are equivalent (i.e., γ satisfies (*) if and only if it satisfies (**)).

b. Exactly one vector $\gamma \in W$ satisfies (**) – i.e., a solution of the minimisation problem exists and is unique.

c. $||f||^2 = ||\gamma||^2 + ||f - \gamma||^2.$

Proof of (10).

a. ($\Rightarrow$) Choose $h \in W$. For all real x, $\gamma + xh \in W$ also. Therefore, if (*) holds, then (setting $\delta = f - \gamma$)

$$(||f - (\gamma + xh)||^2 \geq ||f - \gamma||^2 \Rightarrow ||\delta||^2 - 2x(\delta, h) + x^2 ||h||^2 \geq ||\delta||^2$$
$$\Rightarrow x^2 ||h||^2 - 2x(\delta, h) \geq 0) \ \forall x \in \mathbb{R}.$$

This is possible only if $(\delta, h) = 0$. Thus (**) holds.
($\Leftarrow$) If (**) holds then we have

$$((f - \gamma) \perp (\gamma - h) \overset{(9)}{\Rightarrow} ||f - h||^2 = ||f - \gamma||^2 + ||\gamma - h||^2$$
$$\Rightarrow ||f - h||^2 \geq ||f - \gamma||^2) \ \forall h \in W$$

Thus (*) holds.

b. Suppose that both γ_1 and γ_2 are solutions to (**) in W. Since $\gamma_1 - \gamma_2 \in W$, $(f - \gamma_1) \perp (\gamma_1 - \gamma_2)$ and hence, by (9),

$$||f - \gamma_2||^2 = ||f - \gamma_1||^2 + ||\gamma_1 - \gamma_2||^2.$$

By (a), however, γ_1 and γ_2 both also satisfy (*), so

$$||f - \gamma_1||^2 = \min_{g \in W} ||f - g||^2 = ||f - \gamma_2||^2$$

and hence $||\gamma_1 - \gamma_2||^2 = 0 \Rightarrow \gamma_1 = \gamma_2.$

5

c. Since $\gamma \in W$,

$$(f - \gamma) \perp \gamma \overset{(9)}{\Rightarrow} ||f||^2 = ||f - \gamma||^2 + ||\gamma||^2$$

as desired. $\qquad \square$

Definition. We denote by $\pi_W f$ the orthogonal projection of f on W.

Note. $||\pi_W f|| \leq ||f||$, with equality iff $\pi_W f = f$ – i.e., iff $f \in W$. (For, by 10(c), $||f||^2 = ||\pi_W f||^2 + ||f - \pi_W f||^2$.)

It's easy to see that

$$W = \{f \in V : \pi_W f = f\} = \{\pi_W f : f \in V\}.$$

Definition. The ORTHOGONAL COMPLEMENT of W in V is defined to be

$$W^\perp := \{h \in V : h \perp g \ \forall g \in W\}.$$

Note that $W^\perp = \{h \in V : \pi_W h = 0\}$.

11. $W^\perp$ is a subspace of V.

12. $\pi_W : V \to V$ is linear, idempotent and self-adjoint.

   Proof. We abbreviate π_W to π. Let $a_1, a_2 \in \mathbb{R}$ and $f, f_1, f_2 \in V$ be arbitrary. Then we have by (10) that $f_1 - \pi f_1$ and $f_2 - \pi f_2$ are in $W^\perp$ and hence by (11) that

   $$(a_1 f_1 + a_2 f_2) - (a_1 \pi f_1 + a_2 \pi f_2) = a_1(f_1 - \pi f_1) + a_2(f_2 - \pi f_2) \in W^\perp \quad (*)$$

   Since $\pi f_1, \pi f_2 \in W$ and W is a subspace, $a_1 \pi f_1 + a_2 \pi f_2 \in W$; therefore, by (10) and (*) above, $\pi(a_1 f_1 + a_2 f_1) = a_1 \pi f_1 + a_2 \pi f_2$. Thus π is linear. We also have by (10) that $\pi(\pi f) = \pi f$, since $\pi f \in W$; thus π is idempotent.

   Finally, since $\pi f_1, \pi f_2 \in W$, once more by (10) we have that $(f_1 - \pi f_1, \pi f_2) = 0$; thus

   $$(f_1, \pi f_2) = \big(f_1 + (\pi f_1 - \pi f_1), \pi f_2\big) = \big((f_1 - \pi f_1) + \pi f_1, \pi f_2\big)$$
   $$= (f_1 - \pi f_1, \pi f_2) + (\pi f_1, \pi f_2) = (\pi f_1, \pi f_2).$$

   Similarly, $(\pi f_1, f_2) = (\pi f_1, \pi f_2)$, so that $(f_1, \pi f_2) = (\pi f_1, f_2)$. Thus π is self-adjoint. $\qquad \square$

13. We have from the above description of π_W that $W^\perp = \{f - \pi_W f : f \in V\}$.

14. (This is a converse to (12).) If $U : V \to V$ is linear, idempotent and self-adjoint, then U is an orthogonal projection to some subspace (i.e., there is a subspace W' of V so that $U = \pi_{W'}$).

15. Given an arbitrary $f \in V$, we may write uniquely $f = g + h$, with $g \in W$ and $h \in W^\perp$. In fact, $g = \pi_W f$ and $h = \pi_{W^\perp} f$. From this we conclude that $\pi_{W^\perp} \circ \pi_W \equiv 0 \equiv \pi_W \circ \pi_{W^\perp}$ and $(W^\perp)^\perp = W$.

16. Suppose that W_1 and W_2 are two subspaces of V such that $W_2 \subseteq W_1$. Then $\pi_{W_2} f = \pi_{W_2}(\pi_{W_1} f)$ and $\|\pi_{W_2} f\| \leq \|\pi_{W_1} f\|$, with equality iff $\pi_{W_1} f \in W_2$.

Lecture 3

Note. The above concepts and statements (regarding projections etc.) are valid in any Hilbert space, but we are particularly interested in the case $V = L^2(S, \mathcal{A}, P)$.

Note. If V is a Hilbert space and W is a subspace of V, then W is a Hilbert space when equipped with the same inner product as V.

Homework 1

1. If $V = L^2(S, \mathcal{A}, P)$, show that V is finite-dimensional if P is concentrated on a finite number of points in S. You may assume that the one-point sets $\{s\}$ are measurable.

2. Suppose that $S = [0, 1]$, $\mathcal{A}$ is the Borel field (on $[0, 1]$) and P is the uniform probability measure. Let $V = L^2$ and, for I, J fixed disjoint subintervals of S, define

$$W = W_{I,J} := \{f \in V : f = 0 \text{ a.e. on } I \text{ and } f \text{ is constant a.e. on } J\}.$$

   Show that W is a subspace and find $W^\perp$. Also compute $\pi_W f$ for $f \in V$ arbitrary.

3. Let $S = \mathbb{R}^1$, $\mathcal{A} = \mathcal{B}^1$ and P be arbitrary, and set $V = L^2$. Suppose that $s \in V$ is such that $E(e^{ts}) < \infty$ for all t sufficiently small (i.e., for all t in a neighbourhood of 0). Show that the subspace spanned by $\{1, s, s^2, \ldots\}$ is equal to V. (HINT: Check first that the hypothesis implies that $1, s, s^2, \ldots$ are indeed in V. Then check that, if $g \in V$ satisfies $g \perp s^2$ for $r = 0, 1, 2, \ldots$, then $g = 0$ a.e.(P). This may be done by using the uniqueness of the moment-generating function.)

Definition. Let $S = \{s\}$ and $V = L^2(S, \mathcal{A}, P)$. Let $(R, \mathcal{C})$ be a measurable space, and let $F : S \to R$ be a measurable function. If we let $Q = P \circ F^{-1}$ (so that $Q(T) = P(F^{-1}[T])$), then $F(s)$ is called a STATISTIC with corresponding probability space $(R, \mathcal{C}, Q)$. $W = L^2(R, \mathcal{C}, Q)$ is isomorphic to the subspace $\tilde{W} = L^2(S, F^{-1}[\mathcal{C}], P)$ of V.

Application to prediction

Let $S = \mathbb{R}^{k+1}$, $\mathcal{A} = \mathcal{B}^{k+1}$ be the Borel field in $\mathbb{R}^{k+1}$, P be arbitrary and $V = L^2$. Let $s = (X_1, \ldots, X_k; Y)$.

A PREDICTOR of Y is a Borel function $G = G(\underline{X})$ of $\underline{X} = (X_1, \ldots, X_k)$. We assume that $E(Y^2) < \infty$ and take the MSE of G, i.e., $E(|G(\underline{X}) - Y|^2)$, as a criterion. What should we mean by saying that G is the "best" predictor of Y?

i. No restriction on G: Consider the set W of all measurable $G = G(\underline{X})$ with $E(|G|^2) < \infty$. W is clearly (isomorphic to) a subspace of V and, for $G \in W$, $E(G - Y)^2 = ||Y - G||^2$.

Then the *best* predictor of Y is just the orthogonal projection of Y on W, which is the same as the conditional expectation of Y given $\underline{X} = (X_1, \ldots, X_k)$.

Proof (informal). Let $G^*(\underline{X}) = E(Y \mid \underline{X})$. For an arbitrary $G = G(\underline{X}) \in W$,

$$||Y - G||^2 = ||Y - G^*||^2 + ||G - G^*||^2 + 2(Y - G^*, G^* - G),$$

but

$$
\begin{aligned}
(Y - G^*, G^* - G) &= E\big((Y - G^*)(G^* - G)\big) \\
&= E\big[E((Y - G^*)(G^* - G) \mid \underline{X})\big] \\
&= E\big[(G^* - G)E(Y - G^* \mid \underline{X})\big] = 0,
\end{aligned}
$$

so that $||Y - G||^2 = ||Y - G^*||^2 + ||G - G^*||^2$, whence G^* must be the unique projection.

ii. G an affine function: We require that G be an affine function of $\underline{X}$ – i.e., that there be constants $a_0, a_1, \ldots, a_k$ such that $G(\underline{X}) = G(X_1, \ldots, X_k) = a_0 + \sum_{i=1}^k a_i X_i$ for all $\underline{X}$. The class of such G is a subspace W' of the space W defined in the previous case. The best predictor of Y in this class is the orthogonal projection of Y on W', which is called the LINEAR REGRESSION of Y on $(X_1, \ldots, X_k)$.

Lecture 4

We return to predicting Y using an affine function of $\underline{X}$. We define

$$W := \mathrm{Span}\{1, X_1, \ldots, X_k\}$$

and denote by $\hat{Y}$ the orthogonal projection of Y on W. $\hat{Y}$ is characterized by the two facts that

(*) $Y - \hat{Y} \perp 1$, and

(**) $Y - \hat{Y} \perp X_i^0$ for $i = 1, \ldots, k$

where $X_i^0 = X_i - EX_i$. Since $W = \text{Span}\{1, X_1, \ldots, X_k\}$, we may suppose that $\hat{Y} = \beta_0 + \sum_{i=1}^{k} \beta_i X_i^0$. From (*), $\beta_0 = EY$; and, from (**), $\Sigma\beta = \mathbf{c}$ (the 'normal equation'), where $\beta = (\beta_1, \ldots, \beta_k)^T$, $\mathbf{c} = (c_1, \ldots, c_k)^T$, $\Sigma = (\sigma_{ij})$, $c_i = E(Y^0 X_i^0) = \text{Cov}(X_i, Y)$, $\sigma_{ij} = E(X_i^0 X_j^0) = \text{Cov}(X_i, X_j)$ and $Y^0 = Y - EY$. We have (by considering the minimization problem) that there exists a solution β to these two equations; and (by uniqueness of the orthogonal projection) that, if β is any such solution, then $\hat{Y} = \beta_0 + \sum_{i=1}^{k} \beta_i X_i^0$. Σ is positive semi-definite and symmetric.

Homework 1

4. Show that Σ is nonsingular iff, whenever $P(a_1 X_1^0 + \cdots + a_k X_k^0 = 0) = 1$, $a_1 = \cdots = a_k = 0$; and that this is true iff, whenever $P(b_0 + b_1 X_1 + \cdots + b_k X_k = 0) = 1$, $b_0 = b_1 = \cdots = b_k = 0$.

Let us assume that Σ is nonsingular; then $\beta = \Sigma^{-1}\mathbf{c}$ and $\hat{Y} = EY + \sum_{i=1}^{k} \beta_i X_i^0$.

Note.

i. $\hat{Y}$ is called the LINEAR REGRESSION of Y on $(X_1, \ldots, X_k)$, or the AFFINE REGRESSION or the LINEAR REGRESSION of Y on $(1, X_1, \ldots, X_k)$.

ii. $\hat{Y}^0 = \sum_{i=1}^{k} \beta_i X_i^0$ is the projection of Y^0 on $\text{Span}\{X_1^0, \ldots, X_k^0\}$. Thus

$$\text{Var } Y = ||Y^0||^2 = ||Y^0 - \hat{Y}^0||^2 + ||\hat{Y}^0||^2 = \text{Var}(Y - \hat{Y}) + \text{Var } \hat{Y}$$

or, more suggestively, Var(predictand) = Var(residual) + Var(regression).

A related problem concerns

$$R := \sup_{a_1, \ldots, a_k} \text{Corr}(Y, a_1 X_1 + \cdots + a_k X_k) = ?$$

We have that

$$\text{Corr}\left(Y, \sum a_i X_i\right) = \text{Corr}\left(Y^0, \sum a_i X_i^0\right) = \frac{1}{||Y^0|| \, ||L||} \text{Cov}(Y^0, L)$$

$$= \frac{1}{||Y^0|| \, ||L||}(Y^0, L) = \frac{1}{||Y^0||}\left(Y^0, \frac{L}{||L||}\right),$$

where $L = \sum a_i X_i^0$. Since $Y^0 = (Y^0 - \hat{Y}^0) + \hat{Y}^0$,

$$\left(Y^0, \frac{L}{||L||}\right) = \left(\hat{Y}^0, \frac{L}{||L||}\right) \leq ||\hat{Y}^0||$$

with equality iff $\frac{L}{||L||} = d\hat{Y}^0$ for some $d > 0$ (we have used the Cauchy-Schwarz inequality). In particular, $c(\beta_1, \ldots, \beta_k)$ (with c a positive constant) are the maximizing

9

choices of $(a_1, \ldots, a_k)$. Plugging in any one of these maximizing choices gives us that $R = \frac{\|\hat{Y}^0\|}{\|Y^0\|}$ and hence that $R^2 = \frac{\text{Var } \hat{Y}}{\text{Var } Y}$, from which we conclude that

$$(1 - R^2)\text{Var } Y = \text{Var}(Y - \hat{Y}).$$

From the above discussion we see that Hilbert spaces are related to regression, and hence to statistics.

Note. Suppose that $k = 1$, and that we have data

Serial #	
1	(x_1, y_1)
2	(x_2, y_2)
$\vdots$	$\vdots$
n	(x_n, y_n).

We may then let S be the set consisting of the points $(1; x_1, y_1), \ldots, (n; x_n, y_n)$, to each of which we assign probability $1/n$. If we define $X(i, x_i, y_i) = x_i$ and $Y(i, x_i, y_i) = y_i$ for $i = 1, 2, \ldots, n$, then $EX = \bar{x}$ and $EY = \bar{y}$. $\hat{Y}$ is the affine regression of y on x and R is the correlation between x and y, which is

$$\frac{1}{S_x S_y}\left[\left(\sum x_i y_i\right) - n\bar{x}\,\bar{y}\right].$$

This extends also to the case $k > 1$.

Lecture 5

Classical estimation problem for inference

In the following, S is a sample space, with sample point s; $\mathcal{A}$ is a σ-field on S; and $\mathcal{P}$ is a set of probability measures P on $\mathcal{A}$, indexed by a set $\Theta = \{\theta\}$. We call Θ the PARAMETER SPACE. (The distinction between probability and statistics is that, in probability, Θ has only one element, whereas, in statistics, Θ is richer.)

Suppose we are given a function $g : \Theta \to \Theta$ and a sample point $s \in S$. We are interested in estimating the actual value of g using s, and describing its quality.

Example 1. Estimate $g(\theta)$ from iid $X_i = \theta + e_i$, where the e_i are iid with distribution symmetric around 0. We let $S = \{X_1, \ldots, X_n\}$ and $\Theta = (-\infty, \infty)$, and define g by $g(\theta) = \theta$ for all $\theta \in \Theta$. We might have:

    a. X_is iid $N(\theta, 1)$.

    b. X_is iid double exponential with density $\frac{1}{2}e^{-|x-\theta|}$ (for $-\infty < x < \infty$), with respect to Lebesgue measure.

10

c. X_is iid Cauchy, with density $\frac{1}{\pi(1+(x-\theta)^2)}$.

Possible estimates are $t_1(s) = \overline{X}$, $t_2(s) = \text{median}\{X_1, \ldots, X_n\}$ and

$$t_3(s) = 10\% \text{ of the trimmed mean in } \{X_1, \ldots, X_n\};$$

there are many others.

In the general case, $(S, \mathcal{A}, P_\theta)$, $\theta \in \Theta$, an ESTIMATE (of $g(\theta)$) is a measurable function t on S such that

$$E_\theta(t^2) = \int_S t(s)^2 dP_\theta(s) < \infty \ \forall \theta \in \Theta.$$

What is a "good" estimate?

Suppose that the loss involved in estimating $g(\theta)$ to be t when it is actually g is $L(t, g)$. (Some important choices of loss functions are $L(t, g) = |t - g|$ – the absolute error – and $L(t, g) = |t - g|^2$ – the square error.) Then the EXPECTED LOSS for a particular estimate t (and $\theta \in \Theta$) is

$$R_t(\theta) = E_\theta\big(L(t(s), g(\theta))\big).$$

R_t is called the RISK FUNCTION for t. For t to be a "good" estimate, we want R_t "small".

We consider now a heuristic for the square error function:

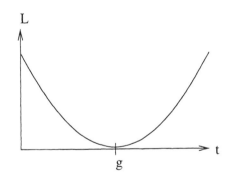

Assume that $L \geq 0$ and that, for each g, $L(g, g) = 0$ and $L(\cdot, g)$ is a smooth function of t. Then

$$L(t, g) = 0 + (t - g)\frac{\partial}{\partial t}L(t, g)\Big|_g + \frac{1}{2}a(g)(t - g)^2 + \cdots = \frac{1}{2}a(g)(t - g)^2 + \cdots$$

where $a(g) \geq 0$. Let us assume that in fact $a(g) > 0$; then we define

$$R_t(\theta) := \frac{1}{2}a(g)E_\theta\big(t(s) - g(\theta)\big)^2,$$

so that R_t is locally proportional to $E_\theta(t - g)^2$, the MSE in t at θ.

Assume henceforth that $R_t(\theta) = E_\theta(t - g)^2$ and denote by $b_t(\theta) = E_\theta(t) - g(\theta)$ the 'bias' of t at θ.

11

1. $R_t(\theta) = \mathrm{Var}_\theta(t) + \big[b_t(\theta)\big]^2$.

Note. It is possible to regard $P_\theta(|t(s) - g(\theta)| > \varepsilon)$ (for $\varepsilon > 0$ small) – i.e., the distribution of t – as a criterion for how "good" the estimate t is. Now, for $Z \geq 0$, $EZ = \int_0^\infty P(Z \geq z)dz$; hence

$$R_t(\theta) = \int_0^\infty P_\theta\big(|t(s) - g(\theta)| > \sqrt{z}\big)dz.$$

There are several approaches to making R_t small. Three of them are:

ADMISSIBILITY: The estimate t is INADMISSIBLE if there is some estimate t' such that $R_{t'}(\theta) \leq R_t(\theta)$ for all $\theta \in \Theta$, and the inequality is strict for at least one θ. t_0 is admissible if it is not inadmissible. (This may be called the "sure-thing principle".)

MINIMAXITY: The estimate t_0 is MINIMAX if

$$\sup_{\theta \in \Theta} R_{t_0}(\theta) \leq \sup_{\theta \in \Theta} R_t(\theta)$$

for all estimates t.

BAYES ESTIMATION: Let λ be a probability on Θ and let $\overline{R}_t = \int_\Theta R_t(\theta)d\lambda$ be the average risk with respect to λ. The estimate t^* is then BAYES (with respect to λ) if $\overline{R}_{t^*} = \inf_t \overline{R}_t$.

2. If t^* has constant risk, i.e., $R_{t^*}(\theta) = c$ for all $\theta \in \Theta$, and t^* is Bayes with respect to some probability λ on Θ, then t^* is minimax.

Proof. Let t be arbitrary; then

$$c = \sup_\theta R_{t^*}(\theta) = \overline{R}_{t^*} \leq \overline{R}_t \leq \sup_\theta R_t(\theta).$$

$\square$

3. If t^* is the essentially unique Bayes estimate with respect to a probability λ on Θ, then t^* is admissible.

Proof. Suppose that t is such that $R_t(\theta) \leq R_{t^*}(\theta)$ for all $\theta \in \Theta$; then $\overline{R}_t \leq \overline{R}_{t^*}$. Hence, by the definition of essential uniqueness,

$$P_\theta(t^* = t) = 1 \;\forall \theta \in \Theta;$$

it follows that $R_{t^*}(\theta) = R_t(\theta)$ for all $\theta \in \Theta$. $\square$

Another approach to making R_t small is:

UNBIASEDNESS: We require all estimates t to be unbiased – i.e., $E_\theta(t) = g(\theta) \Leftrightarrow b_t(\theta) = 0$ for all $\theta \in \Theta$.

Several questions arise:

    i. Are there any unbiased estimates at all?

   ii. If so, which t, if any, has minimum variance at a given θ? (We call such a t a LOCALLY MINIMUM-VARIANCE UNBIASED ESTIMATE.)

  iii. If there is a locally minimum variance unbiased estimate, is it independent of θ? (If so, then it is the uniformly minimum-variance unbiased estimate. If this estimate exists, what is it?)

    There are two approaches: (I) general; and (II) sufficiency (i.e., via complete sufficient statistics).

Chapter 2

Lecture 6

Bayes estimation

We have the setup from the previous lecture: $(S, \mathcal{A}, P_\theta)$, $\theta \in \Theta$. We want to estimate $g(\theta)$. Let $\mathcal{B}$ be a σ-field in Θ and λ a probability on $\mathcal{B}$. We assume that g is $\mathcal{B}$-measurable and that $g \in L^2(\Theta, \mathcal{B}, \lambda)$ – i.e., that $\int_\Theta g(\theta)^2 d\lambda(\theta) < \infty$. We regard θ as a random element and $P_\theta(A)$ as a conditional probability that $s \in A$, given θ. Let $w = (s, \theta)$, $\Omega = S \times \Theta$ and $\mathcal{C} = \mathcal{A} \times \mathcal{B}$ be the smallest σ-field containing all sets of the form $A \times B$ with $A \in \mathcal{A}$ and $B \in \mathcal{B}$. We assume that $\theta \mapsto P_\theta(A)$ is $\mathcal{B}$-measurable for all $A \in \mathcal{A}$.

Lemma 1. *There exists a unique probability measure M on $\mathcal{C}$ such that*

$$M(A \times B) = \int_B P_\theta(A) d\lambda(\theta) \ \forall A \in \mathcal{A}, B \in \mathcal{B}.$$

(λ is the distribution of θ, P_θ is the distribution of s given θ and M is the joint distribution of $w = (s, \theta)$. We will see the explicit formula for Q_s – the distribution of θ given s – soon.)

Consider an estimate t. Our assumption on g is that

$$E(g^2) = \int_\Omega g^2 dM = \int_\Theta g^2 d\lambda < \infty,$$

so

$$\overline{R}_t = \int_\Theta R_t(\theta) d\lambda(\theta) = \int_\Theta E_\theta \big(t - g(\theta)\big)^2 d\lambda(\theta)$$

$$= E\big(E((t - g)^2 \mid \theta)\big) = E(t - g)^2 = ||t - g||^2$$

(the norm taken in $L^2(\Omega, \mathcal{C}, M)$). We would like to choose a t to minimize this quantity. Since t is a function only of s, the desired minimizing estimate – which we will denote by t^* – is the projection of g to the subspace of all $\mathcal{A}$-measurable functions $t(s)$ satisfying $E_M\big(t(s)^2\big) < \infty$. We know that $t^*(s) = E\big(g(\theta) \mid s\big)$.

4.   a. There exists a minimizing t^* (this is the Bayes estimate for g with respect to λ).

     b. The minimizing t^* is unique up to $\overline{P}$-equivalence, where

$$\overline{P}(A) = \int_\Theta P_\theta(A) d\lambda(\theta)$$

     for $A \in \mathcal{A}$.

     c.

$$\overline{R}_{t^*} = \inf_t \overline{R}_t = ||t^* - g||^2 = ||g||^2 - ||t^*||^2$$

$$= \int_\Omega g^2 dM - \int_\Omega (t^*)^2 dM = \int_\Theta g(\theta)^2 d\lambda(\theta) - \int_S t^*(s)^2 d\overline{P}(s).$$

     d. $t^*(s) = E\big(g(\theta) \mid s\big)$, where (s, θ) is distributed according to M.

     Proof. Clear.   □

Note. $\overline{P}$ is the marginal distribution of s.

Some explicit formulas

Suppose we begin with a dominated family $\{P_\theta : \theta \in \Theta\}$ – i.e., a family such that there exists a σ-finite μ such that each P_θ admits a density, say ℓ_θ with respect to μ such that ℓ_θ is measurable, $0 \leq \ell_\theta(s) < \infty$ and

$$P_\theta(A) = \int_A \ell_\theta(s) d\mu(s) \ \forall A \in \mathcal{A}.$$

Two basic examples are:

i. $S = \{s_1, s_2, \ldots\}$ is countable and μ is counting measure. Then $\ell_\theta(s)$ is the P_θ-probability of the atomic set $\{s\}$.

ii. $S = \mathbb{R}^k$, $\mathcal{A} = \mathcal{B}^k$ and μ is Lebesgue measure on $\mathbb{R}^k$. Then ℓ_θ is the probability density in the familiar sense.

We assume that $(s, \theta) \mapsto \ell_\theta(s)$ is a $\mathcal{C}$-measurable function. We write $\nu = \mu \times \lambda$.

5.   a. M is given by

$$\frac{dM}{d\nu}(s, \theta) = \ell_\theta(s).$$

     b. $\overline{P}$ is given by

$$\frac{d\overline{P}}{d\mu}(s) = \int_\Theta \ell_\theta(s) d\lambda(\theta).$$

15

c. For all $s \in S$, Q_s, the conditional probability on Θ given s when (s, θ) is distributed according to M, is well-defined and given by

$$\frac{dQ_s}{d\lambda}(\theta) = \frac{\ell_\theta(s)}{\bar{\ell}(s)},$$

where $\bar{\ell}(s) = \frac{d\bar{P}}{d\mu}$.

d. For all $g \in L^2$,

$$E(g \mid s) = t^*(s) = \int_\Theta g(\theta) dQ_s(\theta) = \int_\Theta g(\theta) \frac{\ell_\theta(s)}{\bar{\ell}(s)} d\lambda(\theta).$$

Proof. These are all easy consequences of Fubini's theorem. (For example, since

$$\Pr(s \in A, \theta \in B) = E(I_A(s) I_B(\theta)) = E(I_B(\theta) E(I_A(s) \mid \theta))$$
$$= E(I_B(\theta) P_\theta(A))$$
$$= E\left(I_B(\theta) \int_A \ell_\theta d\mu\right) = \int_\Theta I_B(\theta) \left[\int_A \ell_\theta(s) d\mu(s)\right] d\lambda(\theta) = \int_{A \times B} \ell_\theta(s) d\nu(s, \theta),$$

we have (a).) $\qquad\qquad\square$

Lecture 7

Note. In any $V = L^2(\Omega, \mathcal{C}, M)$, the constant functions form a subspace, which we will denote by $W_c = W_c(\Omega, \mathcal{C}, M)$. The projection of $f \in V$ on W_c is just $E(f)$.

Note. In the present context, with $w = (s, \theta)$ and $\Omega = S \times \Theta$, s is the datum and θ is the unknown parameter. λ is the prior distribution of θ, Q_s is the posterior distribution (after s is observed) of θ, $t^*(s) = E(g \mid s)$ is the posterior mean of $g(\theta)$, $\theta \mapsto \ell_\theta(s)$ is the likelihood function and $\ell_\theta = \frac{dP_\theta}{d\mu}$.

Note. If we do have a λ on hand but no data, then the Bayes estimate is just $Eg = \int_\Theta g(\theta) d\lambda(\theta)$.

6. If t^* is a Bayes estimate of g, then t^* cannot be unbiased, except in trivial cases.

Proof. Suppose that t^* is unbiased and Bayes (with respect to λ). Then, by unbiasedness,

$$(t^*, g) = E(t^* g) = E(E(t^* g \mid \theta)) = E(g(\theta) E(t^* \mid \theta)) = E(g(\theta)^2) = ||g||^2;$$

but also, since t^* is a Bayes estimate,

$$(t^*, g) = E(E(t^* g \mid s)) = E(t^*(s) E(g \mid s)) = ||t^*||^2.$$

From this we conclude that:

$$||t^* - g||^2 = ||g||^2 - ||t^*||^2 = 0 \Leftrightarrow M(\{w : t^*(s) \neq g(\theta)\}) = 0$$

$$\Leftrightarrow \int_{\Theta} P_\theta(t^*(s) \neq g(\theta))d\lambda(\theta) = 0 \Leftrightarrow P_\theta(t^*(s) = g(\theta)) = 1 \text{ a.e.}(\lambda)$$

This last statement, though, holds iff there is an estimate t such that $P_\theta(t(s) = g(\theta)) = 1$ a.e.(λ). Hence, except in the trivial case, t^* cannot be both Bayes and unbiased. $\qquad\square$

Example 1(a). $s = (X_1, X_2, \ldots, X_n)$, $X_i \overset{iid}{\sim} N(\theta, 1)$ and $\Theta = \mathbb{R}^1$. Let μ be Lebesgue measure; then $d\mu = dX_1 \cdots dX_n$ and $\ell_\theta(s) = \prod_{i=1}^{n} \frac{1}{\sqrt{2\pi}} e^{-\frac{1}{2}(X_i - \theta)^2}$. Consider the estimation of $g(\theta) = \theta$. $\overline{X} = \frac{1}{n}\sum_{i=1}^{n} X_i$ is an unbiased estimate of g (in fact, it is the UMVUE of g), so $\overline{X}$ is not Bayes with respect to any λ. We will see later that:

i. $\overline{X}$ is minimax,

ii. $\overline{X}$ is the pointwise limit of Bayes estimates and

iii. $\overline{X}$ is admissible.

Suppose that λ is the $N(0, \sigma^2)$ distribution, so that $d\lambda(\theta) = \frac{1}{\sqrt{2\pi}} e^{-\frac{1}{2\sigma^2}\theta^2} d\theta$. Given s, $\frac{dQ_s}{d\lambda}(\theta)$ is proportional to $\ell_\theta(s)$, i.e., $dQ_s(\theta) = \varphi_1(s)\ell_\theta(s)d\lambda(\theta)$ for some function φ_1 of s.

Now

$$\ell_\theta(s) = \left(\frac{1}{2\pi}\right)^n e^{-\frac{n}{2}(\overline{X} - \theta)^2 - \frac{n}{2}v},$$

where $v = \frac{1}{n}\sum_{i=1}^{n}(X_i - \overline{X})^2$, so that $dQ_s(\theta) = \varphi_1(s)\varphi_2(s)e^{-\frac{n}{2}(\overline{X} - \theta)^2} e^{-\frac{\theta^2}{2\sigma^2}} d\theta$. Let $\sigma^2 = \frac{1}{nh^2}$ for some $h > 0$, so that

$$dQ_s(\theta) = \varphi_3(s)e^{n\overline{X}\theta - \frac{n}{2}h^2\theta^2 - \frac{n}{2}\theta^2} d\theta = \varphi_4(s)e^{-\frac{n(1+h^2)}{2}\left[\theta - \frac{\overline{X}}{1+h^2}\right]^2} d\theta.$$

Thus $Q_s \sim N\left(\frac{\overline{X}}{1+h^2}, \frac{1}{n(1+h^2)}\right)$ (where $h^2 = \left(\frac{1}{n}\right)/\sigma^2$ is the ratio of observation variance to prior variance) and $t^*(s) = E(\theta \mid s) = \frac{\overline{X}}{1+h^2}$. Therefore

$$t^*(s) = \frac{h^2}{1+h^2} \cdot \underbrace{0}_{\dagger} + \frac{1}{1+h^2}\overline{X} = \underbrace{\frac{1/\sigma^2}{n + 1/\sigma^2} \cdot 0 + \frac{n}{n + 1/\sigma^2}\overline{X}}_{\ddagger},$$

where the 0 (labelled $\dagger$ above) arises from the fact that we are dealing with a Bayes estimate with no data; and the terms marked $\ddagger$ represent a weighted average of prior and data mean, with weights proportional to inverse variance.

We have

$$R_{\overline{X}}(\theta) = \text{Var}_\theta(\overline{X}) + 0 = \frac{1}{n},$$

17

$$R_{t^*}(\theta) = \mathrm{Var}_\theta(t^*) + \left(\frac{\theta}{1+h^2} - \theta\right)^2 = \frac{1}{(1+h^2)^2}\frac{1}{n} + \left(\frac{h^2}{1+h^2}\right)^2 \theta^2$$

and

$$\overline{R}_{t^*} = \frac{1}{(1+h^2)^2}\frac{1}{n} + \left(\frac{h^2}{1+h^2}\right)^2 \sigma^2 = \frac{1}{n(1+h^2)}.$$

(The last can be checked by noting that $\overline{R}_{t^*} = ||\theta - t^*||^2 = ||\theta||^2 - ||t^*||^2$.)

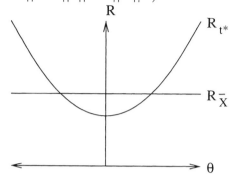

Lecture 8

In the framework $(S, \mathcal{A}, P_\theta)$, $\theta \in \Theta$, set up in the previous lectures, suppose that $dP_\theta = \ell_\theta d\mu$ on S, where $d\mu$ is a fixed measure (this is another way of saying that ℓ_θ is the likelihood function). Given s, suppose $\hat{\theta} = \hat{\theta}(s)$ is the point in Θ such that $\ell_{\hat{\theta}(s)}(s) = \sup_{\theta \in \Theta} \ell_\theta(s)$. Then $\hat{\theta}$ is the (or an) ML estimate of θ.

Given a function g on Θ, $g(\hat{\theta}(s))$ is the ML estimate of $g(\theta)$.

Homework 2

1. What is $\hat{\theta}$ in Example 1(b) (on page 10)? In Example 1(c) (no explicit answer is available in this case)? Assume that $\Theta = \mathbb{R}^1$.

We return to our investigation of Example 1(a).

Example 1. Here μ is n-dimensional Lebesgue measure and $\ell_\theta(s) = \varphi(s)e^{-\frac{n}{2}(\overline{X}-\theta)^2}$, so that $\hat{\theta}(s) = \overline{X}$. The ML estimate of θ^2 is $\overline{X}^2$, the ML estimate of $|\theta|$ is $|\overline{X}|$, etc.

Under square error loss, we know that, if λ is distributed as $N(0, \sigma^2)$ with $\sigma^2 = \frac{1}{nh^2}$, then the Bayes estimate of θ is $t^*(s) = \frac{\overline{X}}{1+h^2}$. The Bayes estimate of θ^2 is

$$\int_{\mathbb{R}^1} \theta^2 dQ_s(\theta) = \frac{1}{n(1+h^2)} + \frac{\overline{X}^2}{(1+h^2)^2}$$

and the Bayes estimate of $|\theta|$ is $\int_{\mathbb{R}^1} |\theta| dQ_s(\theta)$, which is a sum involving special functions.

Note.

18

i. We have seen that the Bayes estimates of θ converge to $\overline{X}$, which is ML and UMVUE, as $h \to 0$ (i.e., $\sigma^2 \to +\infty$). This does not always happen, however: The Bayes estimates of θ^2 converge to $\overline{X}^2 + \frac{1}{n}$; the ML estimate of θ^2 is $\overline{X}^2$ and the UMVUE of θ^2 is $\overline{X}^2 - \frac{1}{n}$. These three are not identical, but they are close if n is large.

ii. The case $h = 0$ ($\sigma^2 = +\infty$) corresponds to uniform prior ignorance about θ; i.e., for two intervals (a, b) and (c, d) in $\mathbb{R}^1 = \Theta$, $\frac{\lambda(a,b)}{\lambda(c,d)} \to \frac{b-a}{d-c}$ as $h \downarrow 0$ (from Homework 2).

iii. $t^*(s) = \frac{\overline{X}}{1+h^2}$ (h fixed) is admissible because t^* is an essentially unique Bayes estimate in $L^2(M)$ (i.e., if t_0 is also Bayes, then $\overline{P}(t_0 = t^*) = 1$, which is equivalent to saying that $P_\theta(t_0 = t^*) = 1$ for all $\theta \in \Theta$). It follows that, for any constant c, $t(s) = \alpha \overline{X} + (1 - \alpha)c$ is admissible for any $0 \le \alpha \le 1$. (Let $X_i' = X_i + c$ and $\theta' = \theta + c$, and apply the above result.)

Homework 2

2. Find a necessary and sufficient condition such that $b_0 + b_1 X_1 + \cdots + b_n X_n$ be admissible for θ.

For any $h > 0$, $R_{t^*}(\theta) = u + v\theta^2$, so $\sup_{\theta \in \Theta} R_{t^*}(\theta) = +\infty$ and t^* is not minimax. We do have, however, that:

i. $\overline{X}$ is minimax.

Proof. Choose any estimate t.

$$\sup_\theta R_t(\theta) \ge \overline{R}_t \ge \overline{R}_{t^*} = \frac{1}{n(1+h^2)};$$

so, since h is arbitrary, $\sup_\theta R_t(\theta) \ge \frac{1}{n}$, which is the (constant) risk of $\overline{X}$. Thus $\overline{X}$ is minimax. $\qquad\square$

ii. $\overline{X}$ is admissible.

Proof. Let t be an estimate such that

$$R_t(\theta) \le R_{\overline{X}}(\theta) = \frac{1}{n} \ \forall \theta \in \Theta$$

and let, for $h > 0$, λ be distributed as $N(0, 1/nh^2)$. We have that $\overline{R}_t \ge \overline{R}_{t^*}$ because t^* is Bayes for λ.

19

Now $t - \theta = (t - t^*) + (t^* - \theta)$ and $(t - t^*) \perp (t^* - \theta)$, so

$$||t - t^*||^2 = ||t - \theta||^2 - ||t^* - \theta||^2 = \overline{R}_t - \overline{R}_{t^*}$$

$$\leq \frac{1}{n} - \overline{R}_{t^*} = \frac{1}{n} - \frac{1}{n(1 + h^2)} = \frac{h^2}{n(1 + h^2)}.$$

We have also that

$$||t - t^*||^2 = \int_{\mathbb{R}^{n+1}} (t - t^*)^2 dM = \int_{\mathbb{R}^n} (t - t^*)^2 d\overline{P}$$

$$= \int_{\mathbb{R}^n} \left(t(s) - t^*(s)\right)^2 \int_\Theta \frac{\sqrt{n}h}{\sqrt{2\pi}} e^{-\frac{nh^2}{2}\theta^2} \ell_\theta(s) d\theta \, d\mu(s)$$

$$= \int_{\mathbb{R}^{n+1}} \left(t(s) - t^*(s)\right)^2 \frac{\sqrt{n}h}{\sqrt{2\pi}} e^{-\frac{nh^2}{2}\theta^2} \ell_\theta(s) d\theta \, d\mu(s).$$

From these two equalities we conclude that

$$\int_{\mathbb{R}^{n+1}} \left(t(s) - t^*(s)\right)^2 \sqrt{\frac{n}{2\pi}} e^{-\frac{nh^2}{2}\theta^2} \ell_\theta(s) d\mu(s) d\theta \leq \frac{h}{n(1 + h^2)}.$$

Letting $h \to 0$ and using Fatou's lemma and the fact that $t^* \to \overline{X}$ (as $h \to 0$), we have that

$$\int_{\mathbb{R}^{n+1}} \left(t(s) - \overline{X}\right)^2 \ell_\theta(s) d\mu(s) d\theta = 0$$

and hence that $\left(t(s) - \overline{X}\right)^2 \ell_\theta(s) = 0$ a.e. (with respect to Lebesgue measure) in $\mathbb{R}^{n+1}$. Since $\ell_\theta(s) > 0$ for all $(s, \theta) \in \mathbb{R}^{n+1}$ by presumption, we have that

$(t(s) - \overline{X} = 0$ a.e. (with respect to Lebesgue measure) on $\mathbb{R}^{n+1})$

$\Rightarrow (t(s) = \overline{X}$ a.e. (with respect to Lebesgue measure) on $S = \mathbb{R}^n)$

$\Rightarrow \left(P_\theta\left(t(s) = \overline{X}\right) = 1 \; \forall \theta \in \Theta\right) \Rightarrow \left(R_t(\theta) = R_{\overline{X}}(\theta) = \frac{1}{n} \; \forall \theta \in \Theta\right).$

$\square$

Lecture 9

Homework 2

3. In Example 1(a), what is the marginal distribution $\overline{P}$ of s? Are the X_is normal and independent under $\overline{P}$? What is the distribution of $\overline{X}$ under $\overline{P}$? Here the prior is assumed to be $\lambda \sim N(0, 1/nh^2)$.

Example 2(a). Let n be a fixed positive integer and $s = (X_1, \ldots, X_n)$ with the X_i iid random variables assuming the values 1 and 0 with probabilities θ and $1 - \theta$ respectively and $\Theta = [0, 1]$. Let μ be counting measure on S, the set of 2^n possible values assumed by s. Then

$$\ell_\theta(s) = P_\theta(\{s\}) = \theta^{T(s)}(1 - \theta)^{n - T(s)}$$

where $T(s) = \sum_{i=1}^n X_i$. The ML estimate is $\hat{\theta}(s) = T(s)/n = \overline{X}$, which is unbiased for θ. (We will see that in fact $\overline{X}$ is the UMVUE.) $R_{\overline{X}}(\theta) = \mathrm{Var}_\theta(\overline{X}) = \theta(1 - \theta)/n$. The ML estimate of $\theta(1-\theta)/n$ is just $\overline{X}(1-\overline{X})/n$, which is *not* unbiased. We shall see later that $\frac{1}{n}\left(\frac{T}{n} - \frac{T}{n}\frac{T-1}{n-1}\right)$ is the UMVUE for $\theta(1-\theta)/n$. Let λ be the Beta distribution $B(a, b)$, i.e.,

$$d\lambda(\theta) = \frac{\theta^{a-1}(1 - \theta)^{b-1}}{B(a, b)}d\theta$$

for $0 \leq \theta \leq 1$, with parameters $a, b > 0$. Here $B(a, b) = \frac{\Gamma(a)\Gamma(b)}{\Gamma(a+b)}$ is the Beta function; it is easy to check that $E_\lambda(\theta) = \frac{a}{a+b}$ and $\mathrm{Var}_\lambda(\theta) = \frac{ab}{(a+b)^2(a+b+1)}$.

We visualize the product space $\Omega = S \times \Theta$ as a unit interval attached to each point of S:

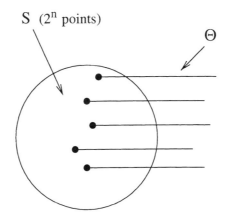

S (2ⁿ points)

Θ

We let μ be counting measure on S, λ prior measure on Θ and $\nu = \mu \times \lambda$ the product measure on Ω. We have $dM(s, \theta) = \ell_\theta(s)d\nu(s, \theta)$ on Ω, so that $M(C) = \int_C \ell_\theta(s)d\nu(s, \theta)$ for all $C \in \mathcal{C} = \mathcal{A} \times \mathcal{B}$, and

$$dQ_s(\theta) = \frac{\ell_\theta(s)}{\overline{\ell}(s)}d\lambda(\theta) = \varphi_1(s)\ell_\theta(s)d\lambda(\theta)$$
$$= \varphi_2(s)\theta^{T(s)}(1 - \theta)^{n-T(s)}d\lambda(\theta)$$
$$= \varphi_2'(s)\theta^{a+T(s)-1}(1 - \theta)^{b+n-T(s)-1}d\theta,$$

where $\varphi_2'(s) = B(a+T(s), b+n-T(s))^{-1}$, so that Q_s is the $B(a+T(s), b+n-T(s))$ distribution.

21

The Bayes estimate for θ is $t^*(s) = E_{Q_s}(\theta) = \frac{a+T(s)}{a+b+n}$; similarly, the Bayes estimate for $\theta(1-\theta)$ is $E_{Q_s}(\theta) - E_{Q_s}(\theta^2)$.

$$R_{t^*}(\theta) = \mathrm{Var}(t^*) + \left[b_{t^*}(\theta)\right]^2 = \frac{n\theta(1-\theta)}{(a+b+n)^2} + \left[\frac{a+n\theta}{a+b+n} - \theta\right]^2.$$

If we choose $a = \frac{\sqrt{n}}{2} = b$, then t^* becomes $t_m^* = \frac{T+\sqrt{n}/2}{n+\sqrt{n}}$ and

$$R_{t_m^*}(\theta) = \frac{n}{4(\sqrt{n}+n)^2} < \frac{1}{4n}.$$

Hence t_m^* is a Bayes estimate with constant risk; therefore it must be minimax. The graphs for the risk functions of $\overline{X}$ and t_m^* look like:

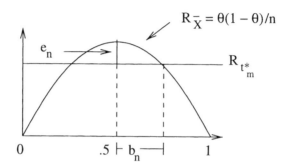

As $n \to \infty$, $b_n \to 0$ and $\frac{e_n}{1/(4n)} \to 0$. Neither $\overline{X}$ nor t_m^* is perfect; for example, if $n = 100$ and $T = 0$, then $\overline{X} = \hat{\theta} = 0$, which is too low, but $t_m^* = \frac{0+5}{100+10} \approx 4\frac{1}{2}\%$, which may be too high.

Note. With $a = \frac{\sqrt{n}}{2} = b$, the prior mean is $\frac{1}{2}$ and the prior variance is $\frac{n/4}{n(\sqrt{n}+1)} = \frac{1}{4(\sqrt{n}+1)}$.

Homework 2

4. Show that $\overline{X} = \frac{T(s)}{n}$ is admissible in two ways:

   a. Show that $\overline{X}$ is the pointwise limit of t^* as $a \downarrow 0$ and $b \downarrow 0$.

   b. Redefine the loss function by $L(t,\theta) = \frac{(t-\theta)^2}{\theta(1-\theta)}$, so that $R_t(\theta) = \frac{E_\theta(t-\theta)^2}{\theta(1-\theta)}$. (Admissibility with respect to this loss function is equivalent to admissibility with respect to the loss function $L(t,\theta) = (t-\theta)^2$.)

5. Show that $\overline{X}$ is the unique Bayes estimate with respect to some λ.

22

Lecture 10

Example 2(b) (negative binomial sampling). We have $\theta \in \Theta = (0,1)$. We choose a positive integer k and observe the iid

$$X_i = \begin{cases} 1 & \text{with probability } \theta \\ 0 & \text{with probability } 1 - \theta \end{cases}$$

until exactly k 1's are observed.

Let N be the total number of X_is observed. Here $s = (X_1, \ldots, X_N)$ and N is a random variable.

Now let S be the set of all possible values of s; then S is countable (obviously $N \geq k$. There are $\binom{r-1}{k-1}$ values of s corresponding to $N = r$). Let μ be counting measure on S (such that sample points with N the same have the same probability). Then

$$\ell_\theta(s) = P_\theta(\text{observing } s) = \theta^{k-1}(1-\theta)^{(N(s)-1)-(k-1)}\theta = \theta^k(1-\theta)^{N(s)-k}.$$

The MLE of θ is $\overline{X} = \frac{k}{N(s)}$, which is not unbiased. Note that $N = N_1 + \cdots + N_k$, where N_1 is the number of trials until the first 'success' (i.e., observation of a 1), N_2 is the number of additional trials required for the second success etc. The N_is are iid, so that $E_\theta(N) = kE_\theta(N_1)$ and $\text{Var}_\theta(N) = k\,\text{Var}_\theta(N_1)$. Since $P_\theta(N_1 = r) = (1-\theta)^{r-1}\theta$ (for $r = 1, 2, \ldots$), we have $E_\theta(N_1) = 1/\theta$ and $\text{Var}_\theta(N_1) = (1-\theta)/\theta^2$. Thus

$$E_\theta\left(\frac{N}{k}\right) = \frac{1}{k}E_\theta(N) = E_\theta(N_1) = \frac{1}{\theta};$$

remember, however, that, by the Cauchy-Schwarz inequality, $E(X)E(1/X) \geq 1$ for any random variable $X > 0$, with equality iff $P(X = c) = 1$, and so

$$E_\theta\left(\frac{k}{N}\right) > \frac{1}{1/\theta} = \theta$$

– i.e., $\overline{X} = \frac{k}{N(s)}$ is biased upwards.

It can be shown by the Rao-Blackwell theorem that the estimate $t = \frac{k-1}{N-1}$ is unbiased when $k \geq 2$. In fact, t is (by the Lehmann-Scheffé theorem or a geometrical approach) the UMVUE. (Heuristically, we see that, if $s = (X_1, \ldots, X_{N-1}, X_N)$, then necessarily $X_N \equiv 1$ (we stop as soon as we observe the kth 1) and so only $(X_1, \ldots, X_{N-1})$ constitute the active part. Then $t(s) = \frac{\text{number of successes in active part}}{\text{number of trials in active part}}$.) To see that t is unbiased, note that

$$P_\theta(N = r) = \binom{r-1}{k-1}\theta^k(1-\theta)^{r-k}$$

for $r = k, k+1, \ldots$, so that

$$E_\theta(t) = \sum_{r=k}^{\infty} \frac{k-1}{r-1}P_\theta(N = r) = \theta.$$

23

We have $\overline{X} - t(s) = \frac{k}{N(s)} - \frac{k-1}{N(s)-1} = \frac{N(s)-k}{N(s)(N(s)-1)} \geq 0$, the inequality being strict with positive probability; so $E_\theta(\overline{X}) > E_\theta(t) = \theta$.

Bayes estimates

Let λ be a prior probability measure on $(0,1)$. As always $dQ_s(\theta) = \varphi_1(s)\ell_\theta(s)d\lambda(\theta)$. Since $\ell_\theta(s)$ is as in Example 2(a), formally the Bayes estimate here is identical to the one there. In particular, $\frac{a+k}{a+b+N(s)}$ is admissible and Bayes with respect to the $B(a,b)$ prior with $a, b > 0$.

Note. Although the MLEs in Examples 2(a) and 2(b) are formally identical, the risk functions are different. In Example 2(b), $R_{\overline{X}}(\theta) = \mathrm{Var}_\theta(\overline{X}) + \left[E_\theta(\overline{X} - \theta)\right]^2$ and $R_t(\theta) = \mathrm{Var}_\theta(t)$ are complicated expressions.

Example 2(c). Depicted here are the stopping points for Examples 2(a) and 2(b), along with those of another possible (two-stage) sampling scheme:

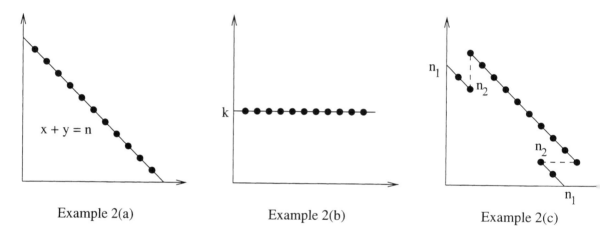

Example 2(a)　　　　Example 2(b)　　　　Example 2(c)

Here (as in any scheme) the likelihood function is $\ell_\theta(s) = \theta^{T(s)}(1-\theta)^{N(s)-T(s)}$, where $T(s)$ is the number of successes (and, of course, $N(s) - T(s)$ is the number of failures). μ is counting measure and the MLE is $\hat{\theta}(s) = \frac{T(s)}{N(s)}$ always.

$s = (X_1, \ldots, X_N)$, where $N = n_1$ or $N = n_1 + n_2$ depending on s. How do we estimate θ? What is the precision of this estimate?

Chapter 3

Lecture 11

Unbiasedness has an appealing property, which we discuss here: Choose any estimate $t(s)$. Imagining for the moment that s is unknown but θ is provided, what is the best predictor for t?

Let λ be the prior; this determines M, as above. Regard t and g as elements of $L^2(M)$.

7. (t is an unbiased estimate of g) $\Leftrightarrow$ (for any choice of a probability λ on θ, g is the best (in MSE) predictor for t).

   Proof. If t is an unbiased estimate of g, then, for any λ, $E(t \mid \theta) = g$ — i.e., g is the projection of t to the subspace of functions in $L^2(M)$ which depend only on θ; or, equivalently, g is the best predictor of t in the sense of $||\cdot||_M$. Conversely, assume that each one-point set in Θ is measurable and take λ to be degenerate at a point θ. The assumption that g is the best predictor of t tells us that $g(\theta) = E(t \mid \theta)$ or, equivalently, that t is an unbiased estimate of g. $\qquad\square$

Unbiased estimation; likelihood ratio

Choose and fix a $\theta \in \Theta$ and let $\delta \in \Theta$. Assume that P_δ is absolutely continuous with respect to P_θ on $\mathcal{A}$; then, by the Radon-Nikodym theorem, there exists an $\mathcal{A}$-measurable function $\Omega_{\delta,\theta}$ satisfying $0 \leq \Omega_{\delta,\theta} \leq +\infty$ and $dP_\delta = \Omega_{\delta,\theta} dP_\theta$ (i.e., $P_\delta(A) = \int_A \Omega_{\delta,\theta}(s) dP_\theta(s)$ for all $A \in \mathcal{A}$).

Note. Suppose that we begin with $dP_\delta(\theta) = \ell_\delta(s) d\mu(s)$ on S, where μ is given, and that we know that $P_\theta(A) = 0 \Rightarrow P_\delta(A) = 0$ (i.e., that P_δ is absolutely continuous with respect to P_θ). Then

$$\Omega_{\delta,\theta}(s) = \begin{cases} \ell_\delta(s)/\ell_\theta(s) & \text{if } 0 < \ell_\theta(s) < \infty \\ 1 & \text{if } \ell_\theta(s) = 0 \end{cases}$$

is an explicit formula for the likelihood ratio. In fact $\Omega_{\delta,\theta}$ can be defined arbitrarily on the set $\{s : \ell_\theta(s) = 0\}$.

In estimating g on the basis of s, let U_g be the class of all unbiased estimates of g. For an estimate $t \in U_g$, the risk function is given by $R_t(\theta) = E_\theta(t - g)^2 = \mathrm{Var}_\theta(t)$. Two questions arise immediately: What is the infimum (over U_g) of the variances at a given θ of the various estimates to g? Is it attained?

Remember that we fixed a $\theta \in \Theta$ above. Let $V_\theta = L^2(S, \mathcal{A}, P_\theta)$; then we assume throughout that

$$\{\Omega_{\delta,\theta} : \delta \in \Theta\} \subseteq V_\theta,$$

i.e., that $E_\theta(\Omega_{\delta,\theta}^2) < +\infty$. Let W_θ be the subspace of V_θ spanned by $\{\Omega_{\delta,\theta} : \delta \in \Theta\}$.

8.   a. U_g is non-empty iff $U_g \cap W_\theta$ is non-empty.

We assume henceforth that U_g is non-empty. Then:

b. $U_g \cap W_\theta$ contains (essentially) only *one* estimate $\tilde{t}$.

c. $\tilde{t}$ is the orthogonal projection on W_θ of every $t \in U_g$.

d. $\mathrm{Var}_\theta(t) \geq \mathrm{Var}_\theta(\tilde{t})$ for all $t \in U_g$.

Note. The above means that $\tilde{t} \in U_g \cap W_\theta$ is the LMVUE of g at θ. $\tilde{t}$ often depends on θ, and this is the problem in practice.

Proof of (8). Note first that

1. $1 \in W_\theta$ (since $\Omega_{\theta,\theta} \equiv 1$).
2. For any t, $E_\delta(t) = \int_S t(s)dP_\delta(s) = \int_S t(s)\Omega_{\delta,\theta}(s)dP_\theta(s)$, so that $E_\delta(t) = (t, \Omega_{\delta,\theta})_\theta$, where $(\cdot, \cdot)_\theta$ is the inner product in $L^2(S, \mathcal{B}, P_\theta)$.

To prove (a), suppose that U_g is non-empty. Let $t \in U_g$ and define $\tilde{t} = \pi t$, where $\pi = \pi_{W_\theta}$ is the orthogonal projection on W_θ. Then, for any $\delta \in \Theta$,

$$E_\delta(\tilde{t}) = (\tilde{t}, \Omega_{\delta,\theta})_\theta = (\pi t, \Omega_{\delta,\theta})_\theta = (t, \pi\Omega_{\delta,\theta})_\theta = (t, \Omega_{\delta,\theta})_\theta = E_\delta(t) = g(\delta).$$

To prove (b), suppose $t_1, t_2 \in U_g \cap W_\theta$; then

$$(t_1 - t_2, \Omega_{\delta,\theta})_\theta = E_\delta(t_1 - t_2) = g(\delta) - g(\delta) = 0 \; \forall \delta \in \Theta.$$

Hence $(t_1 - t_2) \perp \Omega_{\delta,\theta}$ for all $\delta \in \Theta$, and so $(t_1 - t_2) \perp W_\theta$; but $t_1 - t_2 \in W_\theta$, so

$$(t_1 - t_2) \perp (t_1 - t_2) \Rightarrow t_1 - t_2 = 0 \Rightarrow P_\theta(t_1 = t_2) = 1.$$

It follows by absolute continuity that $P_\delta(t_1 = t_2) = 1$ for all $\delta \in \Theta$.

(c) follows from (b) and the above construction.

(d) follows from (c) since $\tilde{t}$ is unbiased for g.   □

Note. In verifying (8), please remember that, if $E_\delta(t) = g(\delta) = E_\delta(\tilde{t})$ for all $\delta \in \Theta$, then $\mathrm{Var}_\theta(t) = E_\theta(t^2) - g(\theta)^2$ and $\mathrm{Var}_\theta(\tilde{t}^2) = E_\theta(\tilde{t}^2) - g(\theta)^2$, so that $\mathrm{Var}_\theta(\tilde{t}) \leq \mathrm{Var}_\theta(t)$, with equality iff $t = \tilde{t}$.

Lecture 12

We may restate (8) as follows:

8'.   a. For some $\tilde{t} \in W_\theta$, $E_\delta(\tilde{t}) = E_\delta(t)$ for all $\delta \in \Theta$ and $t \in U_g$.

b. $\pi_{W_\theta} t$ is such a $\tilde{t}$, and is the (essentially) unique such.

c. We have that

$$R_{\tilde{t}}(\theta) = E_\theta\big(\tilde{t} - g(\theta)\big)^2 = \mathrm{Var}_\theta(\tilde{t}) + \big[b_t(\theta)\big]^2 \leq \mathrm{Var}_\theta(t) + \big[b_t(\theta)\big]^2 = R_t(\theta)$$

with equality iff $t = \tilde{t}$.

d. $\tilde{t}$ is (essentially) the only unbiased estimate of g which belongs to W_θ.

9.   a. An estimate t is the locally MVUE of $g(\delta) := E_\delta(t)$ at θ iff t has finite variance at each δ and $t \in W_\theta$.

b. An estimate t is the UMVUE of $g(\theta) := E_\theta(t)$ iff $t \in \bigcap_{\theta \in \Theta} W_\theta$ (we assume that $\Omega_{\delta,\theta} \in L^2(P_\theta)$ for all $\theta, \delta \in \Theta$).

9(b) above raises the question: Can we describe $C := \bigcap_{\theta \in \Theta} W_\theta$? We know it contains the constant functions; does it contain any others?

10 (Lehman-Scheffé). Write

$$\tilde{V} = \bigcap_{\theta \in \Theta} V_\theta \cap \{v : E_\delta(v) = 0 \ \forall \delta \in \Theta\}.$$

If t has finite variance for each δ (i.e., $t \in \bigcap_{\theta \in \Theta} V_\theta$), then $t \in C$ iff, for each $\delta \in \Theta$, we have

$$E_\delta(tu) = 0 \ \forall u \in \tilde{V}.$$

Proof. Suppose that $t \in C$. Then $t \perp_\delta W_\delta^\perp$ for all $\delta \in \Theta$. Now, for all $u \in \tilde{V}$, u is an unbiased estimate of 0; from (8), we know that 0 is the projection of u to any W_δ. Since $u = 0 + u$, we must therefore have $u \in W_\delta^\perp$, so that $t \perp_\delta u$ for each δ – i.e., $E_\delta(tu) = 0$ for all δ.

Conversely, fix a $\theta \in \Theta$ and write $t = \pi t + u$, where $u = t - \pi t$ and $\pi = \pi_{W_\theta}$. Then $E_\delta(u) = 0$ for all δ and hence, by hypothesis, we have that

$$E_\theta(u^2) + E_\theta(u \cdot \pi t) = E_\theta\big((\pi t + u)u\big) = E_\theta(tu) = 0$$
$$\Rightarrow E_\theta(u^2) = -E_\theta(u \cdot \pi t) = -(\pi t, u) = 0$$

– i.e., $u = 0$ a.e.(P_θ) and hence, by absolute continuity of each P_δ, $u = 0$ a.e.(P_δ) also for every $\delta \in \Theta$. This means that $t = \pi t = \pi_{W_\theta} t \Rightarrow t \in W_\theta$; since $\theta \in \Theta$ was arbitrary, this means that $t \in \bigcap_{\theta \in \Theta} W_\theta = C$ as desired. $\square$

Example 1(d). We have $s = (X_1, \ldots, X_n)$, with the X_i iid as $N(\theta, 1)$, and $\Theta = \{1, 2\}$. We have explicitly that

$$\ell_\theta(s) \propto e^{-\frac{n}{2}(\overline{X}-\theta)^2}$$

and

$$\Omega_{\delta,\theta}(s) = e^{n(\delta-\theta)\overline{X} - \frac{n}{2}(\delta^2-\theta^2)}.$$

Choose $\theta = 1$; then

$$W_\theta = \mathrm{Span}\{\Omega_{11}, \Omega_{21}\} = \mathrm{Span}\{1, e^{n\overline{X}}\} = \{a + be^{n\overline{X}} : a, b \in \mathbb{R}\}.$$

Let $g(\delta) = \delta$. Since $\overline{X}$ is an unbiased estimate of g, we have a unique unbiased estimate of g in W_θ. Hence we want

$$
\begin{aligned}
E_1(a + be^{n\overline{X}}) &= 1 \\
E_2(a + be^{n\overline{X}}) &= 2
\end{aligned}
\tag{*}
$$

Since $\sqrt{n}(\overline{X} - \delta) \sim N(0, 1)$ for $\delta \in \Theta$, under δ, using the MGF of $N(0, 1)$, we have

$$E_\delta(e^{n\overline{X}}) = e^{n\delta} E_\delta(e^{\sqrt{n}\cdot\sqrt{n}(\overline{X}-\delta)}) = e^{n\delta + \frac{1}{2}n}$$

for any $\delta \in \Theta$. Solving (*), we find a and b ($b > 0$). Thus $a + be^{n\overline{X}}$ is LMVU for $E_\theta(X_1)$ at $\theta = 1$. This is not, however, a reasonable estimate. We already know that $\Theta = \{1, 2\}$, but this estimate takes values in $(-\infty, \infty)$. (Since Θ is not connected, we don't have Taylor's theorem here. Also, the LMVUE at $\theta = 2$ is a very different function of $\overline{X}$.) This is absurd. MSE is not suitable because g takes on only two values.

We try changing our parameter space to $\Theta = (\ell, u)$. Now

$$W_\theta = \mathrm{Span}\{\Omega_{\delta,\theta} : \ell < \delta < u\} = \mathrm{Span}\{e^{t\overline{X}} : t \text{ is sufficiently small}\}$$

(in the last set, 't is sufficiently small' means 'for t in a fixed neighbourhood of 0'). It can be shown that

$$W_\theta = \{f(\overline{X}) : f \text{ is a Borel function and } E_\theta f^2 < +\infty\}.$$

Proof (outline). Since $\frac{e^{t_1\overline{X}} - e^{t_2\overline{X}}}{t_1 - t_2} \in W_\theta$, we have that $\frac{d}{dt}e^{t\overline{X}} \in W_\theta$. Hence $\overline{X}e^{t\overline{X}} \in W_\theta$ for $|t|$ sufficiently small. Iterating this reasoning gives us that $\overline{X}^2 e^{t\overline{X}}, \ldots, \overline{X}^i e^{t\overline{X}}, \ldots \in W_\theta$ for $|t|$ sufficiently small (what "sufficiently small" means depends on i). Thus $1, \overline{X}, \overline{X}^2, \ldots \in W_\theta$ and hence W_θ is as desired.

Since $\overline{X}$ is an unbiased estimate of $E_\delta(X_1)$ which belongs to W_θ, $\overline{X}$ is LMVU at θ, and hence $\overline{X}$ is the UMVUE; since $\overline{X}^2 - \frac{1}{n}$ is an unbiased estimate of $[E_\delta(X_1)]^2$, $\overline{X}^2 - \frac{1}{n}$ is the UMVUE for $[E_\delta(X_1)]^2$. (Here W_θ essentially does not depend on θ, and

$$C = \{f(\overline{X}) : f \text{ is Borel and } E_\theta f^2 < +\infty \; \forall \theta \in \Theta\}$$

28

by our above computation.)

Let $A \subseteq S$ be such that $P_1(A) \neq P_2(A)$(for example, $A = \{s : X_1(s) > 3/2\}$). Then $a + bI_A$ is an unbiased estimate of θ if a and b are chosen properly. Indeed, there are many unbiased estimates. To find the "best", we try to minimize variances, noting that

$$W_1 = \text{Span}\{\Omega_{11}, \Omega_{21}\} = \{a + be^{n\overline{X}} : a, b \in \mathbb{R}\}$$

is the class of all estimates which are unbiased for their own expected values and have minimum variance when $\theta = 1$ and hence that there is a $t_1 \in W_1$ such that $E_\delta(t_1) = \delta$ for $\delta = 1, 2$. (*Exercise*: What is t_1?).

Similarly, $W_2 = \{a + be^{-n\overline{X}} : a, b \in \mathbb{R}\}$ and there is a $t_2 \in W_2$ such that $E_\delta(t_2) = \delta$ for $\delta = 1, 2$. (*Exercise*: What is t_2?) $t_1 \neq t_2$, however; in fact, C is the set of all UMVUEs, which is just the set of constant functions.

As noted, the Neyman-Pearson theory implies that we should use $a + bI_A$ with $A = \{s : \overline{X} > c\}$ and $b > 0$. We should also restrict the estimation theory to a continuum of values (i.e., should have only connected Θ).

Chapter 4

Lecture 13

The score function, Fisher information and bounds

Let Θ be an open interval in $\mathbb{R}^1$ and suppose that $dP_\theta(s) = \ell_\theta(s)d\mu(s)$, where μ is a fixed measure on S. Suppose that $\theta \mapsto \ell_\theta(s)$ is differentiable for each fixed s; then $\delta \mapsto \Omega_{\delta,\theta}(s) = \frac{\ell_\delta(s)}{\ell_\theta(s)}$ is also differentiable for each fixed (s,θ). If we use dashes for derivatives with respect to the parameters as described, then

$$\Omega'_{\theta,\theta}(s) = \frac{\ell'_\theta(s)}{\ell_\theta(s)} =: \gamma_\theta^{(1)}(s)$$

is the SCORE FUNCTION at θ (given s). We also define $I(\theta) := E_\theta\big(\gamma_\theta^{(1)}(s)\big)^2$, the FISHER INFORMATION (for estimating θ) in s.

Note.

$$\Big(\int_S \ell_\delta(s)d\mu(s) = 1 \ \forall \delta \in \Theta\Big)$$

$$\Rightarrow \Big(\int_S \Omega'_{\delta,\theta}(s)dP_\theta(s) = \int_S \frac{\ell'_\delta(s)}{\ell_\theta(s)}\ell_\theta(s)d\mu(s) = \int_S \ell'_\delta(s)d\mu(s) = 0 \ \forall \delta \in \Theta\Big)$$

$$\Rightarrow E_\theta\big(\gamma_\theta^{(1)}(s)\big) = E_\theta\big(\Omega'_{\theta,\theta}(s)\big) = 0 \Rightarrow I(\theta) = \mathrm{Var}_\theta(\gamma_\theta^{(1)})$$

Similarly, we have $\int_S \ell''_\delta(s)d\mu(s) = 0$, $\int_S \ell'''_\delta(s)d\mu(s) = 0$, etc. for all $\delta \in \Theta$, so that $E_\theta\big(\gamma_\theta^{(j)}(s)\big) = 0$ for $j = 1, 2, 3, \ldots$, where $\gamma_\theta^{(j)}(s) = \big(\frac{\partial^j \ell_\theta(s)}{\partial \theta^j}\big)/\ell_\theta(s)$. Conditions under which the interchanging of differentiation and integration (as above) is valid will be given later.

Suppose that we are interested in W_θ and want some concrete method of constructing it. We have that

$$\Omega_{\delta,\theta}(s) = \Omega_{\theta,\theta} + (\delta - \theta)\gamma_\theta^{(1)}(s) + \frac{1}{2}(\delta - \theta)^2 \gamma_\theta^{(2)}(s) + \cdots,$$

which suggests that $W_\theta = \mathrm{Span}\{1, \gamma_\theta^{(1)}, \gamma_\theta^{(2)}, \ldots\}$. We will see that this equality holds exactly in a one-parameter exponential family and approximately in general in large samples. To see that $\gamma_\theta^{(j)} \in W_\theta$, we reason as follows: First, of course, we note that $1 \in W_\theta$. Then, since $\Omega_{\delta,\theta}, \Omega_{\theta,\theta} \in W_\theta$, we have that $\frac{1}{\delta-\theta}(\Omega_{\delta,\theta} - \Omega_{\theta,\theta}) \in W_\theta$ for $\delta \neq \theta$, from which it follows that $\gamma_\theta^{(1)} \in W_\theta$. Similar inductive reasoning allows us to conclude that each $\gamma_\theta^{(j)}$ is in W_θ.

It is clear that 1 and $\gamma_\theta^{(1)}$ are the most important generators if s is very informative, for then only δ near the true θ are important. In any case,

$$W_\theta^{(k)} := \mathrm{Span}\{1, \gamma_\theta^{(1)}, \gamma_\theta^{(2)}, \ldots, \gamma_\theta^{(k)}\} \subseteq W_\theta.$$

We know that, in $V_\theta = L^2(P_\theta)$, every $t \in U_g$ projects to the same $\tilde{t} \in W_\theta$; thus every $t \in U_g$ has the same projection to $W_\theta^{(k)}$ – say $t_{\theta,k}^*$. Then we have:

11. BHATTACHARYA BOUNDS: For each $t \in U_g$,

$$\mathrm{Var}_\theta(t) \geq E_\theta(t_{\theta,k}^*)^2 - \left[g(\theta)\right]^2$$

for $k = 1, 2, \ldots$.

Proof. This follows since

$$\mathrm{Var}_\theta(t) + \left[g(\theta)\right]^2 = E_\theta(t^2) \geq E_\theta(t_{\theta,k}^*)^2.$$

$\square$

Let us consider the case $k = 1$ – i.e., projection to $\mathrm{Span}\{1, \gamma_\theta^{(1)}\}$. We have seen that $1 \perp \gamma_\theta^{(1)}$ – i.e., that $E_\theta(\gamma_\theta^{(1)}) = 0$ – and that $||\gamma_\theta^{(1)}||^2 = I(\theta)$. Hence $\{1, \gamma_\theta^{(1)}/||\gamma_\theta^{(1)}||\}$ is an orthonormal basis in $W_\theta^{(1)}$ and, for any $t \in V_\theta$, the projection $t_{\theta,1}^*$ of t to $W_\theta^{(1)}$ is

$$t_{\theta,1}^* = (1,t)1 + \left(\frac{\gamma_\theta^{(1)}}{||\gamma_\theta^{(1)}||}, t\right)\frac{\gamma_\theta^{(1)}}{||\gamma_\theta^{(1)}||}.$$

Now $(1,t) = E_\theta(t) = g(\theta)$ since t is unbiased, and

$$(\gamma_\theta^{(1)}, t) = E_\theta(t \cdot \gamma_\theta^{(1)}) = \int_S t(s)\frac{\ell_\theta'(s)}{\ell_\theta(s)}dP_\theta(s) = \int_S t(s)\ell_\theta'(s)d\mu(s)$$

$$\overset{(?)}{=} \frac{d}{d\theta}\int_S t(s)\ell_\theta(s)d\mu(s) = \frac{d}{d\theta}g(\theta) = g'(\theta).$$

The above calculations give us that

$$t_{\theta,1}^* = g(\theta) + \frac{g'(\theta)}{||\gamma_\theta^{(1)}(s)||}\frac{\gamma_\theta^{(1)}(s)}{||\gamma_\theta^{(1)}(s)||};$$

since the summands are orthogonal,

$$||t_{\theta,1}^*||^2 = g(\theta)^2 + \frac{(g'(\theta))^2}{||\gamma_\theta^{(1)}(s)||^2} = g(\theta)^2 + \frac{(g'(\theta))^2}{I(\theta)}.$$

From this we see:

12 (Fisher-Darmois-Cramér-Rao). INFORMATION INEQUALITY: For $t \in U_g$,

$$\mathrm{Var}_\theta(t) \geq \frac{(g'(\theta))^2}{I(\theta)}.$$

The Fisher information can be related to the second derivative of the log-likelihood: Let $L_\theta(s) = \log_e \ell_\theta(s)$. Then $L'_\theta(s) = \frac{\ell'_\theta(s)}{\ell_\theta(s)} = \gamma_\theta^{(1)}(s)$ and

$$L''_\theta(s) = \frac{\ell''_\theta(s)}{\ell_\theta(s)} - \left(\frac{\ell'_\theta(s)}{\ell_\theta(s)}\right)^2 = \frac{\ell''_\theta(s)}{\ell_\theta(s)} - [\gamma_\theta^{(1)}]^2;$$

but $E_\theta\big(\ell''_\theta(s)/\ell_\theta(s)\big) = \int_S \ell''_\theta(s)d\mu(s) = 0$, and so

13. $E_\theta\big(L''_\theta(s)\big) = -I(\theta)$.

Exact conditions under which statements (11)–(13) hold are deferred until Lecture 5.1.

Lecture 14

Heuristics for maximum likelihood estimate:

  i. $W_\theta = \mathrm{Span}\{1, \gamma_\theta^{(1)}, \gamma_\theta^{(2)}, \ldots\}$.

 ii. $W_\theta \approx \mathrm{Span}\{1, \gamma_\theta^{(1)}\}$ if s is highly informative.

iii. The MLE $\hat{\theta}(s) \dot{\in} W_\theta$ (whatever θ may be!).

The last item gives us that:

 iv. $\hat{\theta}$ is approximately the UMVUE of its own expected value function (the same is true of estimates related to $\hat{\theta}$ in certain ways).

Let $\hat{\theta}(s)$ be the MLE of θ and assume that $\hat{\theta}$ is close to θ. Since $\hat{\theta}(s)$ maximizes L_δ, we have

$$0 = L'_{\hat{\theta}} = L'_\theta + (\hat{\theta} - \theta)L''_\theta + \cdots \approx L'_\theta + (\hat{\theta} - \theta)L''_\theta.$$

Assume also that the experiment (that is, $(S, \mathcal{A}, P_\theta)$, $\theta \in \Theta$) is highly informative in the sense that $I(\theta)$ is large (for a given θ). We know that $E_\theta(L'_\theta) = 0$ and $\mathrm{Var}_\theta(L'_\theta) = I(\theta)$; hence, informally, L'_θ is "about" 0, "give or take" about $\sqrt{I(\theta)}$. From (13), $E_\theta(-L''_\theta) = I(\theta)$ – i.e., $E_\theta\big(-\frac{L''_\theta}{I(\theta)}\big) = 1$. Assume that the random variable $-\frac{L''_\theta}{I(\theta)} \approx 1$. Then

$$\hat{\theta} \approx \theta - \frac{L'_\theta}{L''_\theta} = \theta + \frac{L'_\theta}{\sqrt{I(\theta)}} \frac{1}{\sqrt{I(\theta)}} \frac{1}{-L''_\theta/I(\theta)} \approx \theta + \frac{1}{\sqrt{I_\theta}} \frac{\gamma_\theta^{(1)}}{||\gamma_\theta^{(1)}||}, \qquad (*)$$

and hence $\hat\theta$ is nearly in $W_\theta^{(1)} \subseteq W_\theta$; so $\hat\theta$ is nearly LMVU, and hence $\hat\theta$ is nearly the UMVUE (of θ). From (*),

$$E_\theta(\hat\theta) \approx \theta \quad \text{and} \quad \mathrm{Var}_\theta(\hat\theta) \approx \frac{1}{I(\theta)}.$$

The MLE of $g(\theta)$ is $g(\hat\theta)$. Assuming that g is continuously differentiable, we have

$$g(\hat\theta) \approx g(\theta) + g'(\theta)(\hat\theta - \theta).$$

So $g(\hat\theta)$ is nearly in W_θ (since $1 \in W_\theta$ and $\hat\theta$ is nearly in W_θ). Hence

$$E_\theta g(\hat\theta) \approx g(\theta) \quad \text{and} \quad \mathrm{Var}_\theta\big(g(\hat\theta)\big) \approx \frac{(g'(\theta))^2}{I(\theta)},$$

where $\frac{[g'(\theta)]^2}{I(\theta)}$ is the lower bound in (12).

Note. $\frac{I(\theta)}{[g'(\theta)]^2}$ is the information in s for estimating $g(\theta)$.

Suppose that $(S_1, \mathcal{A}_1, P_\theta^{(1)})$ and $(S_2, \mathcal{A}_2, P_\theta^{(2)})$, $\theta \in \Theta$, are independent experiments concerning θ, with sample points s_1 and s_2. Let $s = (s_1, s_2)$, $\mathcal{A} = \mathcal{A}_1 \times \mathcal{A}_2$ and $P_\theta = P_\theta^{(1)} \times P_\theta^{(2)}$, and let $I_i(\theta)$ be the information in s_i for estimating θ ($i = 1, 2$). Then the information in s for estimating θ is $I(\theta) = I_1(\theta) + I_2(\theta)$. (This result extends inductively to any finite number of independent experiments.)

Proof. $dP_\theta^{(i)}(s) = \ell_\theta^{(i)}(s_i)d\mu^{(i)}(s_i)$ for $i = 1, 2$, so $dP_\theta(s) = \ell_\theta^{(1)}(s_1)\ell_\theta^{(2)}(s_2)d\nu(s)$ and hence

$$L_\theta(s) = \log \ell_\theta^{(1)}(s_1) + \log \ell_\theta^{(2)}(s_2) = L_\theta^{(1)}(s) + L_\theta^{(2)}(s).$$

The result now follows from (13). □

Example 1(a). $s = (X_1, \ldots, X_n)$, $X_i \overset{\text{iid}}{\sim} N(\theta, 1)$. The information in s for estimating θ is the sum of the information in $X_1, \ldots, X_n$, respectively, for estimating θ, which sum is (since the X_i are iid) n times the information in X_1, which product is (since X_1 is distributed as $N(\theta, 1)$) just n. $L_\theta'(X_1) = X_1 - \theta = \gamma_\theta^{(1)}(X_1)$ and $\mathrm{Var}_\theta(\gamma_\theta^{(1)}) = 1 = I_1(\theta)$. (We check that $\frac{L_\theta'(s)}{\sqrt{I(\theta)}}$ is about 0, give or take about 1; and $\frac{L_\theta''(s)}{I(\theta)} \approx 1$ (indeed, here it is identically 1).)

Example 2. $X_1, \ldots, X_n, \ldots$ are iid as

$$\begin{cases} 0 & \text{with probability } 1 - \theta \\ 1 & \text{with probability } \theta, \end{cases}$$

and $\Theta = (0, 1)$. $s = (X_1, \ldots, X_N)$, N the stopping time. The three cases we discussed are:

a. $N \equiv n$ (n a fixed positive integer).

33

b. N is the first time k successes (i.e., 1s) are recorded (k a fixed positive integer).

c. Two-stage scheme.

In all cases (even other than (a)–(c) above), $\ell_\theta(s) = \theta^{T(s)}(1-\theta)^{N(s)-T(s)}$, where $T(s) = \sum_{i=1}^{N(s)} X_i$, and hence

$$L'_\theta(s) = \frac{T(s)}{\theta} - \frac{V(s)}{1-\theta}$$

and

$$L''_\theta(s) = -\frac{T(s)}{\theta^2} - \frac{V(s)}{(1-\theta)^2},$$

where $V(s) = N(s) - T(s)$. Since $E_\theta(L'_\theta) = 0$, we have

$$\frac{E_\theta(T)}{\theta} = \frac{E_\theta(V)}{1-\theta}, \tag{4.1}$$

$$I(\theta) = \mathrm{Var}_\theta(L'_\theta) = \frac{\mathrm{Var}_\theta(T)}{\theta^2} + \frac{\mathrm{Var}_\theta(V)}{(1-\theta)^2} - \frac{2}{\theta(1-\theta)}\mathrm{Cov}_\theta(T, V) \tag{4.2}$$

and

$$I(\theta) = E_\theta(-L''_\theta) = \frac{1}{\theta^2}E_\theta(T) + \frac{1}{(1-\theta)^2}E_\theta(V). \tag{4.3}$$

Exercise: What happens in case (a) (i.e., $N \equiv n$)? This is like Example 1(a) except that $I(\theta)^{-1} = \frac{\theta(1-\theta)}{n}$ depends on θ.

Suppose now that we are in case (b). Then $T \equiv k$ and $V(s) = N(s) - k$. Hence, from (4.1), $\frac{k}{\theta} = \frac{E_\theta(N-k)}{1-\theta}$ and therefore $E_\theta(N) = \frac{k}{\theta}$ (which we could also compute directly) and

$$I(\theta) = \frac{k}{\theta^2} + \frac{1}{(1-\theta)^2}E_\theta(N - k) = \frac{k}{\theta^2(1-\theta)}$$

(from equation (4.3) above). Hence the heuristics apply when k is large. In all cases $\hat\theta(s)$ is $\frac{T(s)}{N(s)}$.

Exercise: Derive $\mathrm{Var}_\theta(N)$ from (4.2) and check the behavior of $L'_\theta/\sqrt{I(\theta)}$ and $L''_\theta/\sqrt{I(\theta)}$.

Example 1(e). $s = (X_1, \cdots, X_n)$, with the X_i iid with density $ae^{-b(x-\theta)^4}$ ($a, b > 0$).

Homework 3

1. a. Find a and b such that $\mathrm{Var}_\theta(X_1) = 1$ (to make it comparable to Example 1(a)).

    b. Find $I(\theta)$.

    c. $E_\theta(\overline{X}) \equiv \theta$, so $\overline{X}$ is unbiased for θ. Is $\overline{X}$ the UMVUE? (Note the answer is no.)

    d. What is the UMVUE?

    e. Give an explicit method for finding $\hat\theta(s)$.

Lecture 15

13(a). Suppose $t \in U_g$ is such that

$$\mathrm{Var}_\theta(t) = \frac{[g'(\theta)]^2}{I(\theta)} \quad \forall \theta \in \Theta;$$

then $\{P_\theta : \theta \in \Theta\}$ is a one-parameter exponential family with statistic t – i.e.,

$$\frac{dP_\theta}{d\mu}(s) = \ell_\theta(s) = \varphi(s)e^{A(\theta)+B(\theta)t(s)},$$

where A and B are smooth functions; moreover, $g(\theta) = -A'(\theta)/B'(\theta)$.

Proof. By the same argument as used in the proof of (12), we have that $t \in W_\theta^{(1)}$ for all θ – i.e., that

$$t(s) = a(\theta) + b(\theta)L'_\theta(s) \quad \text{a.e.}(P_\theta)$$

for all θ. From this it follows that $L'_\theta(s) = \alpha(\theta) + \beta(\theta)t(s)$ (if $b \equiv 0$ then $\mathrm{Var}_\theta(t) = 0$ for all θ. We rule out this case) and hence that

$$L_\theta(s) = A(\theta) + B(\theta)t(s) + C(s),$$

where $A(\theta) = \int \alpha(\theta)d\theta$. This gives the required form for $\ell_\theta(s)$. Also,

$$0 = E_\theta(L'_\theta) = \alpha(\theta) + \beta(\theta)E_\theta(t) = \alpha(\theta) + \beta(\theta)g(\theta)$$

and so $g(\theta) = -\alpha(\theta)/\beta(\theta) = -A'(\theta)/B'(\theta)$. □

Note. For a near-rigorous proof, see R. A. Wijsman 1973 AS, pp. 538–542, and V. M. Joshi 1976 AS, pp. 998–1002.

Note. The necessary conditions on $\{P_\theta : \theta \in \Theta\}$ and g are also sufficient for the attainment of the C-R bound. We will see this later.

Example 1(a). Since

$$\ell_\theta(s) = \varphi_1(s)e^{-\frac{n}{2}(\overline{X}-\theta)^2} = \varphi_2(s)e^{-\frac{n\theta^2}{2}+(n\theta)\overline{X}},$$

the C-R bound is attained by $\overline{X}$ for estimating $g(\theta) = \theta$. This implies that $\overline{X}$ is LMVU at θ, which in turn implies that it is UMVU. Also, the C-R bound is not attained by any unbiased estimate of any g which is not an affine function of θ. In particular, since $\overline{X}^2 - \frac{1}{n}$ is an unbiased estimate of $g(\theta) = \theta^2$, it does not attain the C-R bound since $\theta^2 \neq -A'(\theta)/B'(\theta)$. We have seen before, however, that $\overline{X}^2 - \frac{1}{n}$ is the UMVUE.

To study the Bhattacharya bounds, note that $\ell'_\theta = \ell_\theta \cdot [-n\theta + n\overline{X}]$ and $\ell''_\theta = \ell_\theta \cdot [-n\theta + n\overline{X}]^2 + \ell_\theta \cdot [-n]$, so that ℓ'_θ/ℓ_θ is affine in $\overline{X}$ and $\ell''_\theta/\ell_\theta$ is quadratic. This implies that

$$W_\theta^{(1)} = \mathrm{Span}\{1, \ell'_\theta/\ell_\theta\} = \mathrm{Span}\{1, \overline{X}\}$$

and

$$W_\theta^{(2)} = \mathrm{Span}\{1, \ell'_\theta/\ell_\theta, \ell''_\theta/\ell_\theta\} = \mathrm{Span}\{1, \overline{X}, \overline{X}^2\},$$

whence $\overline{X}^2 - \frac{1}{n} \in W_\theta^{(2)}$ attains the Bhattacharya bound and is the UMVUE. In fact, W_θ is the space of *all* functions of $\overline{X}$, and hence any function of $\overline{X}$ (but not θ) is the UMVUE of its expectation.

Example 1(b). $s = (X_1, X_2, \ldots)$ are iid from $\frac{1}{2}e^{-|x-\theta|}$ on $\mathbb{R}^1$. Here W_θ is well-defined (i.e., (8)–(10) hold), but (11)–(13) are not applicable since ℓ_θ is not sufficiently smooth. In such a situation, the following is useful.

14 (Chapman-Robbins). Given $(S, \mathcal{A}, P_\theta)$, $\theta \in \Theta$, with Θ an open interval in $\mathbb{R}^1$, if $t \in U_g$ then

$$\mathrm{Var}_\theta(t) \geq \varlimsup_{\delta \to \theta} \left(\frac{g(\delta) - g(\theta)}{\delta - \theta}\right)^2 \bigg/ E_\theta \left(\frac{\Omega_{\delta,\theta} - 1}{\delta - \theta}\right)^2$$

for all θ such that $\Omega_{\delta,\theta} = \frac{dP_\delta}{dP_\theta}$ exists for all δ in a neighborhood of θ.

Proof.

$$E_\delta(t) = g(\delta) \Rightarrow \int_S t \cdot \Omega_{\delta,\theta} dP_\theta = g(\delta) \Rightarrow \int_S t(\Omega_{\delta,\theta} - 1) dP_\theta = g(\delta) - g(\theta).$$

Dividing by $\delta - \theta$, we find that

$$\int t\left(\frac{\Omega_{\delta,\theta} - 1}{\delta - \theta}\right) dP_\theta = \frac{g(\delta) - g(\theta)}{\delta - \theta}$$

$$\Rightarrow \int_S (t - g(\theta))\left(\frac{\Omega_{\delta,\theta} - 1}{\delta - \theta}\right) dP_\theta = \frac{g(\delta) - g(\theta)}{\delta - \theta}$$

$$\Rightarrow \left(\frac{g(\delta) - g(\theta)}{\delta - \theta}\right)^2 \leq \mathrm{Var}_\theta(t) \cdot E_\theta \left(\frac{\Omega_{\delta,\theta} - 1}{\delta - \theta}\right)^2.$$

$\square$

Note. If g is differentiable at θ, then

$$\mathrm{Var}_\theta(t) \geq (g'(\theta))^2 \bigg/ \varlimsup_{\delta \to \theta} E_\theta \left(\frac{\Omega_{\delta,\theta} - 1}{\delta - \theta}\right)^2.$$

If, further, $\Omega_{\delta,\theta}$ is differentiable (see (12E) below for exact conditions), then this is the same as $\frac{[g'(\theta)]^2}{I(\theta)}$.

Homework 3

2. What is the Chapman-Robbins bound for $g(\theta) = \theta$ in Example 1(b)?

3. In Example 1(c), s consists of n iid observations from $\frac{1}{\pi(1+(x-\theta)^2)}$. For any g, the C-R bound is not attained by any t; but $\hat{\theta}$ has nearly the variance $\frac{1}{I(\theta)}$ if $I(\theta)$ is large. Here $I(\theta) = nI_1(\theta)$. Show that $I_1(\theta) = \frac{1}{2}$.

Chapter 5

Lecture 16

Example 1(e). We have X_i iid $ae^{-b(x-\theta)^4}$, with $a, b > 0$ chosen so that this is a density and $\text{Var}_\theta(X_i) = 1$. Then

$$\ell_\theta(s) = \varphi_0(s) \exp\left\{ b\left[\left(4 \sum_{i=1}^{n} X_i^3\right)\theta - 6\left(\sum_{i=1}^{n} X_i^2\right)\theta^2 + 4\left(\sum_{i=1}^{n} X_i\right)\theta^3 \right] + A(\theta) \right\},$$

which is not a one-parameter exponential family. It is called a "curved exponential family".

Sufficient conditions for the Cramér-Rao and Bhattacharya inequalities

As usual, we have $(S, \mathcal{A}, P_\theta)$, $\theta \in \Theta$, where Θ is an open subset of $\mathbb{R}^1$. μ is a fixed measure on S and $dP_\theta(s) = \ell_\theta(s)d\mu(s)$.

Condition 1. $\ell_\theta(s) > 0$ and $\delta \mapsto \ell_\delta(s)$ has, for each $s \in S$, a continuous derivative $\delta \mapsto \ell_\delta'(s)$. Let

$$\gamma_\theta^{(1)}(s) = \frac{\ell_\theta'(s)}{\ell_\theta(s)} = L_\theta'(s).$$

Condition 2. Given any $\theta \in \Theta$, we may find an $\varepsilon = \varepsilon(\theta) > 0$ such that $E_\theta(m_\theta^2) < +\infty$, where

$$m_\theta(s) = \sup_{|\delta - \theta| \leq \varepsilon} |\gamma_\delta^{(1)}(s)|$$

– i.e., $m_\theta \in V_\theta$, which implies that $I(\theta) = E_\theta(\gamma_\theta^{(1)})^2 < +\infty$.

Condition 3. $I(\theta) > 0$.

12E Exact statement of Cramér-Rao inequality: Under conditions 1–3 above, if U_g is non-empty, then g is differentiable and

$$\text{Var}_\theta(t) \geq \frac{(g'(\theta))^2}{I(\theta)} \quad \forall \theta \in \Theta, t \in U_g.$$

38

Proof.

i.
$$\Omega_{\delta,\theta} = \Omega_{\theta,\theta} + (\delta - \theta)\gamma_{\delta^*}^{(1)} = 1 + (\delta - \theta)\gamma_{\delta^*}^{(1)}$$

for some δ^* between θ and δ. By Condition 2, $\Omega_{\delta,\theta} \in V_\theta$ for $|\delta - \theta|$ sufficiently small.

ii.
$$\frac{\Omega_{\delta,\theta}(s) - 1}{\delta - \theta} - \gamma_\theta^{(1)}(s) = \gamma_{\delta^*}^{(1)}(s) - \gamma_\theta^{(1)}(s) \to 0$$

as $\delta \to \theta$ for all $s \in S$. (From Condition 1, $\gamma_\delta^{(1)}$ is continuous.) Also,

$$|\gamma_{\delta^*}^{(1)} - \gamma_\theta^{(1)}| \le 2m_\theta \in V_\theta$$

and hence $E_\theta(\gamma_{\delta^*}^{(1)} - \gamma_\theta^{(1)})^2 \to 0$ as $\delta \to \theta$ (by dominated convergence) – i.e.,

$$\frac{\Omega_{\delta,\theta} - 1}{\delta - \theta} \xrightarrow{V_\theta} \gamma_\theta^{(1)}.$$

From this it follows that $\gamma_\theta^{(1)} \in W_\theta$.

iii. Choose $t \in U_g$. If we let $(\cdot, \cdot)$ and $\|\cdot\|$ be the inner product and norm, respectively, in V_θ, then $E_\delta(t) = E_\theta(t\Omega_{\delta,\theta}) = g(\delta)$ and so

$$(t, \Omega_{\delta,\theta} - 1) = g(\delta) - g(\theta) \Rightarrow (t - g(\theta), \Omega_{\delta,\theta} - 1) = g(\delta) - g(\theta)$$

(since $E_\theta(\Omega_{\delta,\theta} - 1) = 0$), whence

$$\left(t - g(\theta), \frac{\Omega_{\delta,\theta} - 1}{\delta - \theta}\right) = \frac{g(\delta) - g(\theta)}{\delta - \theta} \quad \forall \delta \ne \theta.$$

From (ii) $\frac{g(\delta) - g(\theta)}{\delta - \theta}$ has a finite limit $(t - g(\theta), \gamma_\theta^{(1)})$ as $\delta \to \theta$. Thus g is differentiable and $g'(\theta) = (t - g(\theta), \gamma_\theta^{(1)})$, so that $|g'(\theta)| \le \|t - g(\theta)\| \, \|\gamma_\theta^{(1)}\|$ – i.e., $\text{Var}_\theta(t) \ge \frac{[g'(\theta)]^2}{I(\theta)}$. $\qquad\square$

Note. To know that $\int_S \ell'_\theta d\mu = 0 = \int_S \ell''_\theta d\mu$, it suffices to show that $\delta\ell''_\delta(s)$ exists and is continuous for each s and that

$$\int_S \left\{ \max_{|\delta - \theta| \le \varepsilon} |\ell''_\delta(s)|^2 \right\} d\mu(s) < +\infty$$

for some $\varepsilon = \varepsilon(\theta) > 0$.

Note. Under Conditions 1–3, $\text{Span}\{1, \gamma_\theta^{(1)}\} = W_\theta^{(1)} \subseteq W_\theta$ and $1 \perp \gamma_\theta^{(1)}$ in V_θ. (Take $t \equiv 1$; then $(1, \Omega_{\delta,\theta}) \equiv 1$ and hence

$$\left(1, \frac{\Omega_{\delta,\theta} - 1}{\delta - \theta}\right) = 0 \quad \forall \delta \ne \theta.$$

Letting $\delta \to \theta$, we have that $(1, \gamma_\theta^{(1)}) = 0$.)

39

Let k be a positive integer.

Condition 1_k. For each fixed s, $\theta \mapsto \ell_\theta(s)$ is positive and is k-times continuously differentiable.

Condition 2_k. Given any $\theta \in \Theta$, we may find an $\varepsilon = \varepsilon(\theta) > 0$ such that $E_\theta(m_\theta^2) < +\infty$, where

$$m_\theta(s) = \sup_{|\delta - \theta| \leq \varepsilon} |\gamma_\delta^{(k)}(s)|.$$

(From the above, we have that $1 \perp \gamma_\theta^{(j)}$ for $j = 1, \ldots, k$ – i.e., $E_\theta(\gamma_\theta^{(j)}) = 0$.)

Let $\Sigma_\theta^{(k)}$ be the covariance matrix of $\begin{pmatrix} \gamma_\theta^{(1)} \\ \vdots \\ \gamma_\theta^{(k)} \end{pmatrix}$.

Condition 3_k. $\Sigma_\theta^{(k)}$ is positive definite.

11E. If conditions 1_k–3_k hold and U_g is non-empty, then g is k-times continuously differentiable and

$$\mathrm{Var}_\theta(t) \geq b_k(\theta) \; \forall t \in U_g, \theta \in \Theta,$$

where $b_k(\theta) = h'(\theta)\big[\Sigma_\theta^{(k)}\big]^{-1} h(\theta)$ and $h(\theta) = \begin{pmatrix} g^{(1)}(\theta) \\ \vdots \\ g^{(k)}(\theta) \end{pmatrix}$ (Of course $g^{(j)} = \frac{d^j g}{d\theta^j}$.)

Proof (outline). $1, \gamma_\theta^{(1)}, \ldots, \gamma_\theta^{(k)} \in W_\theta$ and so $W_\theta^{(k)} \subseteq W_\theta$ and

$$\mathrm{Var}_\theta(t) \geq ||t_{\theta,k}^*||^2 - [g(\theta)]^2.$$

Lecture 17

Note.

i. $L_\theta', L_\theta'', \ldots$ are derivatives of $\log_e \ell_\theta$, but $\gamma_\theta^{(1)} = \ell_\theta'/\ell_\theta, \gamma_\theta^{(2)} = \ell_\theta''/\ell_\theta, \ldots$ are *not* the same as $L_\theta', L_\theta'', \ldots$.

ii. Condition 2 in (12E) can be weakened slightly to:

Condition 2'. Given any $\theta \in \Theta$, we may find an $\varepsilon = \varepsilon(\theta) > 0$ such that

$$E\left[\frac{\max_{|\delta - \theta| \leq \varepsilon} |\ell_\delta'(s)|}{\ell_\theta}\right]^2 < +\infty.$$

and condition 2_k in (11E) can be weakened to:

40

Condition $2_k'$. Given any $\theta \in \Theta$, we may find an $\varepsilon = \varepsilon(\theta) > 0$ such that

$$E\left[\frac{\max_{|\delta - \theta| \le \varepsilon} |d^k \ell_\delta(s)/d\delta^k|}{\ell_\theta(s)}\right]^2 < +\infty.$$

iii. Suppose that U_g is non-empty; then (8) implies that the projection of any $t \in U_g$ to W_θ is the (fixed) $\tilde{t} \in U_g \cap W_\theta$. Also, $t^*_{\theta,k}$ is the projection of any $t \in U_g$ to $W_\theta^{(k)} = \text{Span}\{1, \gamma_\theta^{(1)}, \dots, \gamma_\theta^{(k)}\} \subseteq W_\theta$ – i.e., $t^*_{\theta,k}$ is the (affine) "regression" of any $t \in U_g$ on $\{\gamma_\theta^{(1)}, \dots, \gamma_\theta^{(k)}\}$. Thus

$$t^*_{\theta,k} = g(\theta) + \alpha_1 \gamma_\theta^{(1)} + \cdots + \alpha_k \gamma_\theta^{(k)},$$

where $\alpha_1, \dots, \alpha_k$ are determined as in our discussion of regression, and

$$b_k(\theta) = E_\theta(t^*_{\theta,k})^2 - [g(\theta)]^2 = \text{Var}_\theta(\alpha_1 \gamma_\theta^{(1)} + \cdots + \alpha_k \gamma_\theta^{(k)})$$

$$= \left(\frac{dg}{d\theta}, \dots, \frac{d^k g}{d\theta^k}\right) (\Sigma_\theta^{(k)})^{-1} \left(\frac{dg}{d\theta}, \dots, \frac{d^k g}{d\theta^k}\right)'$$

by the regression formula.

iv.
$$b_1(\theta) \le b_2(\theta) \le \cdots \le b_k(\theta) \le \cdots$$

(where $b_1(\theta)$ is the C-R bound) because $W_\theta^{(k)} \subseteq W_\theta^{(k+1)}$. If we define $b(\theta) := \lim_{k \to \infty} b_k(\theta)$, then

$$b(\theta) \le \text{Var}_\theta(\tilde{t}),$$

the actual lower bound at θ for an unbiased estimate of g. We have that $b(\theta) = \text{Var}_\theta(\tilde{t})$ iff $\tilde{t} \in \text{Span}\{1, \gamma_\theta^{(1)}, \gamma_\theta^{(2)}, \dots\}$. This does hold for any g with non-empty U_g *if* the subspace spanned by $\{1, \gamma_\theta^{(1)}, \dots, \gamma_\theta^{(k)}, \dots\}$ is W_θ. This sufficient condition for $b_k \to b$ and $t^*_{\theta,k} \to \tilde{t}$ is plausible since, by the Taylor expansion,

$$\Omega_{\delta,\theta} = 1 + (\delta - \theta)\gamma_\theta^{(1)} + \frac{(\delta - \theta)^2}{2!}\gamma_\theta^{(2)} + \cdots.$$

It holds rigorously in the following case:

15. (One-parameter exponential family) Suppose that

$$\ell_\theta(s) = C(s)e^{A(\theta) + B(\theta)T(s)}$$

where $C(s) > 0$, T is a fixed statistic and B is a continuous strictly monotone function on $\Theta \subseteq \mathbb{R}$; then, under Condition (*) below, we have

 a. $W_\theta^{(k)} = \text{Span}\{1, T, \dots, T^k\}$ for $k = 1, 2, 3, \dots$.
 b. $\text{Span}\{1, T, T^2, \dots\} = W_\theta$ (under θ).

41

c. W_θ is the space of all Borel functions f of T such that $E_\theta\big(f(T)\big)^2 < +\infty$.

d. If U_g is non-empty, then $b_k(\theta) \to b(\theta) = \mathrm{Var}_\theta(\tilde{t})$.

e. $\tilde{t} = E_\theta(t \mid T)$ for all $\theta \in \Theta$ and $t \in U_g$.

f. SUFFICIENCY OF T: Given any $A \subseteq S$, we may find an $h(T)$ independent of θ such that $h(T) = P_\theta(A \mid T)$ for all $\theta \in \Theta$.

Proof. (f) follows from (e) by defining $g(\theta) = P_\theta(A)$ and $t = I_A \in U_g$ and applying (c).

(e) follows from (c) since projection to W_θ is then the same as taking conditional expectation.

(d) follows from (a) and (b) and the above notes.

It now remains only to prove (a)–(c). To this end, let $\xi = B(\delta) - B(\theta)$. Then ξ is the parameter, and takes values in a neighborhood of 0. We have

$$\frac{dP_\xi}{dP_0}(s) = \frac{C(s)e^{A(\delta)+B(\delta)T(s)}}{C(s)e^{A(\theta)+B(\theta)T(s)}} = e^{\xi T(s) - K}.$$

Suppose that

Condition (*). $\xi = B(\delta) - B(\theta)$ takes all values in a neighborhood of 0 as δ varies in a neighborhood of θ.

Under this condition,

$$\int_S e^{\xi T(s)-K} dP_0(s) = \int_S dP_\xi(s) = 1$$

and hence the MGF of T exists for ξ in a neighborhood of 0, and

$$K = K(\xi) = \log_e \int e^{\xi T(s)} dP_0(s)$$

is the cumulant generating function of T under P_θ.

Thus the family of probabilities on S is $\{P_\xi : \xi \text{ in a neighborhood of } 0\}$, where $dP_\xi(s) = e^{\xi T(s) - K(\xi)} dP_0(s)$ – i.e., a one-parameter exponential family with ξ as the "natural" parameter and $T(s)$ as the "natural" statistic. $W_\theta = \mathrm{Span}\{\Omega_{\delta,\theta} : \delta \in \Theta\}$; the spanning set includes $\{e^{\xi T(s)-K(\xi)} : \xi \text{ in a neighbourhood of } 0\}$, so W_θ contains the subspace spanned by $\{e^{\xi T} : \xi \text{ in a neighborhood of } 0\}$. Now

$$\frac{e^{\eta T} - e^{\xi T}}{\eta - \xi} = e^{\xi T}\left(\frac{e^{(\eta-\xi)T} - 1}{\eta - \xi}\right) = e^{\xi T}\frac{(1 + (\eta - \xi)T + \frac{1}{2}(\eta - \xi)^2 T^2 e^{(\eta^* - \xi)T} - 1)}{\eta - \xi}$$

for some η^* between η and ξ. We have, however, that $\frac{1}{2}(\eta - \xi)T^2 e^{(\eta^* - \xi)T} \xrightarrow{L^2} 0$ since the MGFs of T exist around 0. Hence

$$T e^{\xi T} = \lim_{\eta \to \xi} \frac{1}{\eta - \xi}(e^{\eta T} - e^{\xi T}) \in W_\theta.$$

42

Similarly, $T^2 e^{\xi T}, T^3 e^{\xi T}, \ldots$ are in W_θ. Taking $\xi = 0$, we get $\{1, T, T^2, \ldots\} \subseteq W_\theta$, so that the subspace spanned by $\{1, T, T^2, \ldots\}$ is in W_θ; but this subspace is the subspace of all square-integrable Borel functions of T, so $\mathrm{Span}\{1, T, T^2, \ldots\} = W_\theta$ actually, since each $\Omega_{\delta,\theta}$ is a (square-integrable Borel) function of T. $\square$

Example 2. Here $s = (X_1, \ldots, X_N)$, N the total number of trials in a Bernoulli sequence, and $\ell_\theta(s) = \theta^{T(s)}(1-\theta)^{N(s)-T(s)}$, where T, the total number of successes, is $X_1 + X_2 + \cdots + X_N$. In general, this is a curved exponential family.

In Example 2(a), since $N \equiv n$ (a constant),

$$\ell_\theta = e^{n \log_e (1-\theta) + T \log_e (\theta/(1-\theta))},$$

so that T is sufficient and any function of T is the UMVUE of its expected value. $C = \bigcap_{\theta \in \Theta} W_\theta$ is the set of all estimates of the form $f(T)$. The C-R bound b_1 is attained essentially only for $g(\theta) = -A'(\theta)/B'(\theta) = \theta$, i.e., for $g(\theta) = \alpha + \beta\theta$. The kth Bhattacharya bound b_k is attained iff $g(\theta)$ is a polynomial of degree $k \le n$. If $k > n$, then $b_k = b_n = b$.

Lecture 18

Note. In the context of (15), it is sometimes necessary to look at the distribution of the (sufficient) statistic T. Suppose that we have found the distribution function of T for a particular θ – say F_θ; then F_δ is given by

$$dF_\delta(x) = e^{[B(\delta)-B(\theta)]x + [A(\delta)-A(\theta)]} dF_\theta(x),$$

where $x = T(s)$ (so that the distributions of T are a one-parameter exponential family with statistic the identity). (Please check, by computing, that $P_\delta(T \le x) =: F_\delta(x) = \cdots$.)

Example 2(a).

Homework 4

1. U_g is non-empty iff g is a polynomial of degree $\le n$ (in the case of Example 2(a)).

W_θ does not depend on θ; it is the class of all functions of $\overline{X}$, and hence an estimate is a UMVUE of its expected value iff it is a function of $\overline{X}$.

$$\mathrm{Var}_\theta(\overline{X}) = \frac{\theta}{n} - \frac{\theta^2}{n} =: \sigma^2(\theta).$$

We will show that $\sigma^2(\theta)$ has a UMVUE when $n \ge 2$. This UMVUE should be a function of $\overline{X}$. $\frac{\theta}{n}$ may be estimated by $\frac{\overline{X}}{n}$. How about θ^2? Let

$$t = \begin{cases} 1 & \text{if } X_1 \text{ and } X_2 = 1 \\ 0 & \text{otherwise;} \end{cases}$$

43

then $E_\theta t = \theta^2$. We know that the projection to W_θ, which is $E_\theta(t \mid T)$, will give $\tilde t$ for $g(\theta) = \theta^2$. (Taking $E_\theta(t \mid T)$ is called "Blackwellization".)

$$E_\theta(t \mid T = k) = \frac{P_\theta(t = 1, T = k)}{P_\theta(T = k)}$$

$$= \frac{P_\theta(X_1 = 1 = X_2, \text{ exactly } k - 2 \text{ successes in subsequent } n - 2 \text{ trials})}{P_\theta(T = k)}$$

$$= \frac{\theta^2 \binom{n-2}{k-2} \theta^{k-2}(1-\theta)^{n-k}}{\binom{n}{k}\theta^k(1-\theta)^{n-k}} = \frac{\binom{n-2}{k-2}}{\binom{n}{k}} = \frac{k(k-1)}{n(n-1)},$$

which is independent of θ, as expected. Thus

$$\tilde t = \frac{T(T-1)}{n(n-1)},$$

which is the UMVUE of θ^2, and therefore $\sigma^2(\theta)$ may be estimated by

$$\frac{\overline X}{n} - \frac{\overline X}{n}\left(\frac{n\overline X - 1}{n - 1}\right) = \frac{\overline X}{n}\left[1 - \frac{n\overline X - 1}{n - 1}\right],$$

which is a function of $\overline X$ and hence is the UMVUE of $\sigma^2(\theta)$.

Consider the odds ratio $g(\theta) = \frac{\theta}{1-\theta}$. This has no unbiased estimate. Since θ has MLE $\overline X$, $\hat t$, the MLE for this g, is $\frac{\overline X}{1-\overline X}$. Since $P_\theta(\overline X = 1) = \theta^n > 0$, we have $E_\theta(\hat t) = +\infty$, so the expectation breaks down. If, however, $I(\theta) = \frac{n}{\theta(1-\theta)}$ is large – i.e., n is large – then

$$\hat t = \overline X + \cdots + \overline X^n + \frac{\overline X^{n+1}}{1 - \overline X} = \overline X + \cdots + \overline X^n + R_n,$$

where $R_n = \frac{\overline X^{n+1}}{1-\overline X}$. For each $\theta \in (0,1)$, R_n is very small with large probability, and

$$\frac{R_n}{\theta^{n+1}} \to \frac{1}{1 - \theta}$$

in P_θ-probability as $n \to \infty$.

Example 2(b) (Negative binomial sampling). Here

$$\ell_\theta = \theta^k(1-\theta)^{N-k} = \exp\left\{k\log\frac{\theta}{1-\theta} + k\log(1-\theta)\cdot y\right\},$$

where $y = N/k$, so that

$$T = y, \quad A = k\log\big(\theta/(1-\theta)\big) \quad \text{and} \quad B = k\log(1-\theta),$$

and hence $-A'(\theta)/B'(\theta) = 1/\theta$. Thus $E_\theta(y) = 1/\theta$ and $\mathrm{Var}_\theta(y)$ is the C-R bound, and the C-R bound is attained only for $g(\theta) = a + b/\theta$.

44

Now assume $k \geq 3$. We know (even for $k \geq 2$) that $\frac{k-1}{N-1}$ is an unbiased estimate of θ. Since $\frac{k-1}{N-1} = \frac{k-1}{ky-1}$ is a function of y, it is in fact the UMVUE of θ.

Let $\sigma^2(\theta) = \text{Var}_\theta\left(\frac{k-1}{N-1}\right)$. Since $\tilde{t} = \frac{k-1}{N-1}$ is not a polynomial in y – in fact, $\tilde{t} \notin W_{\theta,k} \ \forall k$ – we have (for $g(\theta) = \theta$)

$$b_1(\theta) < b_2(\theta) < \cdots < b_{k+1}(\theta) < \sigma^2(\theta),$$

but $b_k(\theta) \to \sigma^2(\theta)$ as $k \to \infty$. We can, however, find a UMVUE for $\sigma^2(\theta)$ (without knowing what the b_ks are).

Suppose that we can find an unbiased estimate u of θ^2. Then $v = \tilde{t}^2 - u$ is an unbiased estimate of $\sigma^2(\theta)$ ($\sigma^2(\theta) = \text{Var}_\theta(\tilde{t}) = E_\theta(\tilde{t}^2) - \theta^2$).

Let
$$t = \begin{cases} 1 & \text{if } X_1 = 1 = X_2 \\ 0 & \text{otherwise.} \end{cases}$$

Then (even at present) $E_\theta(t) = \theta^2$ and hence $u = E_\theta(t \mid N)$ (the Blackwellization of t) is the UMVUE of θ^2 (when $k \geq 3$).

$$E_\theta(t \mid N = m) = \frac{P_\theta(X_1 = 1 = X_2, N = m)}{P_\theta(N = m)}$$
$$= \frac{\theta^2 \binom{m-3}{k-3} \theta^{k-3}(1-\theta)^{m-k}\theta}{\binom{m-1}{k-1}\theta^{k-1}(1-\theta)^{m-k}\theta} = \frac{\binom{m-3}{k-3}}{\binom{m-1}{k-1}} = \frac{(k-1)(k-2)}{(m-1)(m-2)}$$

– i.e., $u = \frac{(k-1)(k-2)}{(N-1)(N-2)}$ is the UMVUE of θ^2, so that the UMVUE of $\sigma^2(\theta)$ is

$$\left(\frac{k-1}{N-1}\right)^2 - \frac{(k-1)(k-2)}{(N-1)(N-2)} = \frac{(k-1)(N-k)}{(N-1)^2(N-2)}.$$

Homework 4

2. Does every polynomial in θ have an unbiased estimate? (Yes?) Does $\frac{\theta}{1-\theta}$ have an unbiased estimate? (No?)

Chapter 6

Lecture 19

The vector-valued score function and information in the multi-parameter case

Now we have an experiment $(S, \mathcal{A}, P_\theta)$, $\theta = (\theta_1, \ldots, \theta_p) \in \Theta$ with Θ an open set in $\mathbb{R}^p$ and a smooth function $g : \Theta \to \mathbb{R}^1$. We assume that $dP_\theta(s) = \ell_\theta(s)d\mu(s)$ as before, and define $\ell(\theta \mid s) := \ell_\theta(s)$. Assume that ℓ is smooth in θ and let $g_i(\theta) = \frac{\partial}{\partial \theta_i} g(\theta)$, $\ell_i(\theta \mid s) = \frac{\partial}{\partial \theta_i} \ell(\theta \mid s)$ and $\ell_{ij}(\theta \mid s) = \frac{\partial^2}{\partial \theta_i \partial \theta_j} \ell(\theta \mid s)$ for $1 \leq i, j \leq p$. There are two approaches to the present topic in this situation:

Approach 1. Generalize the previous one-dimensional discussion: Suppose that t is unbiased for g – that is to say,

$$\int_S t(s)\ell(\delta \mid s)d\mu(s) = E_\delta(t) = g(\delta)$$

for all $\delta \in \Theta$. Then

$$E_\theta\big(t(s)\ell_i(\theta \mid s)/\ell(\theta \mid s)\big) = \int_S t(s)\ell_i(\theta \mid s)d\mu(s) = g_i(\theta)$$

for $i = 1, \ldots, p$ and hence every $t \in U_g$ has the same projection on $\mathrm{Span}\{1, L_1, \ldots, L_p\}$, where $L(\theta \mid s) = L_\theta(s)$ and

$$L_i(\theta \mid s) = \frac{\partial}{\partial \theta_i} L(\theta \mid s) = \frac{\ell_i(\theta \mid s)}{\ell(\theta \mid s)}.$$

This approach is useful for studies of conditions which ensure that $L_1, L_2, \ldots, L_p$ are in $W_\theta = \mathrm{Span}\{\Omega_{\delta,\theta} : \delta \in \Theta\}$.

Approach 2. Use the result for the θ-real case: Fix $\theta \in \Theta$ and a vector $c = (c_1, \ldots, c_p) \neq 0$, and suppose that δ is restricted to the line passing through θ and $\theta + c$ – in other words, that we consider only $\delta = \theta + \xi c$ for some scalar ξ. (Note that, since Θ is open, if ξ is sufficiently small then $\theta + \xi c \in \Theta$.) Then g becomes a function of ξ for which t remains unbiased. By (12),

$$\mathrm{Var}_\theta(t) \geq [\text{Fisher information in } s \text{ for } g \text{ at } \theta \text{ in the restricted problem}]^{-1}$$

$$= \left(\frac{dg}{d\xi}\bigg|_{\xi=0}\right)^2 \big/ [\text{Fisher information for } \xi \text{ in } s \text{ for estimating } g]$$

Now, since $\delta = \theta + \xi c$,

$$\frac{dg}{d\xi}\bigg|_{\xi=0} = \sum_{i=1}^p \frac{\partial g}{\partial \delta_i}\bigg|_{\delta=\theta} c_i = \sum_{i=1}^p c_i g_i(\theta).$$

The information in the denominator is $E_\theta(dL/d\xi)^2$, and

$$\frac{dL}{d\xi}\bigg|_{\xi=0} = \sum_{i=1}^p c_i L_i(\theta \mid s),$$

so that the information may be expressed explicitly as

$$E_\theta\left(\frac{dL}{d\xi}\right)^2 = \sum_{i=1}^p \sum_{j=1}^p c_i c_j E_\theta\big(L_i(\theta \mid s)L_j(\theta \mid s)\big) = \sum_{i,j} c_i c_j I_{ij},$$

where I_{ij} is the (i, j)th entry of the Fisher information matrix

$$I(\theta) = \big\{\mathrm{Cov}_\theta(L_i(\theta \mid s), L_j(\theta \mid s))\big\}_{p \times p}$$

(where the sample space is S). Let

$$L_{ij} = \frac{\partial L_i}{\partial \theta_j} = \frac{\partial}{\partial \theta_j}\left[\frac{\ell_i}{\ell}\right] = \frac{\ell_{ij}}{\ell} - \frac{\ell_i \ell_j}{\ell^2};$$

then

$$E_\theta(L_{ij}) = \int \ell_{ij}(\theta \mid s)d\mu(s) - E_\theta(L_i L_j) = -E_\theta(L_i L_j)$$

and hence we have the p-dimensional analogue of (13):

13^p. $I(\theta) = \big\{-E_\theta(L_{ij}(\theta \mid s))\big\}$.

The above lower bound for $\mathrm{Var}_\theta(t)$ can now be written as

$$\left[\sum_i c_i g_i(\theta)\right]^2 \bigg/ \left(\sum_{i,j} c_i c_j I_{ij}\right).$$

Let us assume that I is positive definite. It will be shown below that

$$\sup_c \{\text{the bound above}\} = \sum_{i,j} g_i(\theta) I^{ij}(\theta) g_j(\theta), \qquad (*)$$

where $\{I^{ij}(\theta)\} = I^{-1}(\theta)$; and the supremum is achieved when c is a multiple of $h(\theta)I^{-1}(\theta)$, where $h(\theta) = (g_1(\theta), \ldots, g_p(\theta)) = \nabla g(\theta)$.

Thus we have the p-dimensional analogue of (12):

47

12^p. If $t \in U_g$, then $\mathrm{Var}_\theta(t) \geq h(\theta) I^{-1}(\theta) h(\theta)'$.

Assume that this bound is attained, at least approximately; then, for the estimation of g, there exists a one-dimensional problem (namely, the one obtained by restricting δ to $\{\theta + \xi c^* : \xi \in \mathbb{R}\}$, where $c^* = h(\theta) I^{-1}(\theta)$) which is as difficult as the p-dimensional problem.

Proof of (*). For $u = (u_1, \ldots, u_p)$ and $v = (v_1, \ldots, v_p)$ in $\mathbb{R}^p$, let $(u|v) := \sum_{i=1}^p u_i v_i = uv'$ and $||u|| := (u|u)^{1/2}$. Let I be a (fixed) positive definite symmetric $p \times p$ matrix and set $(u|v)_* := \sum_{i,j} u_i I_{ij} v_j = uIv'$ and $||u||_* := (u|u)_*^{1/2}$. Let $g = (g_1, \ldots, g_p)$ be a fixed point in $\mathbb{R}^p$. Consider the maximization over $\underline{a} = (a_1, \ldots, a_p) \in \mathbb{R}^p$ of

$$\frac{(\sum_{i=1}^p a_i g_i)^2}{\sum_{i,j} a_i I_{ij} a_j} = \frac{(\underline{a}g')^2}{||\underline{a}||_*^2} = \frac{(\underline{a}I|gI^{-1})^2}{||\underline{a}||_*^2} = \frac{(\underline{a}|gI^{-1})_*^2}{||\underline{a}||_*^2} = \left(\frac{\underline{a}}{||\underline{a}||_*}\bigg|gI^{-1}\right)_*^2.$$

The unique (up to scalar multiples) maximizing value is given by $\underline{a} = gI^{-1}$ and the maximum value is

$$\left(\frac{gI^{-1}}{||gI^{-1}||_*}\bigg|gI^{-1}\right)_*^2 = \left[\frac{(gI^{-1})I(gI^{-1})'}{||gI^{-1}||_*}\right]^2 = ||gI^{-1}||_*^2 = gI^{-1}g'.$$

$\square$

Lecture 20

We have seen that, with $\theta = (\theta_1, \ldots, \theta_p)$ and fixed g, the "most difficult" one-dimensional problem is with $\delta \in \Theta$ unknown but restricted to

$$\{\theta + \xi c^* : |\xi| \text{ is sufficiently small}\},$$

where $c^* = c^*(\theta) = h(\theta) I^{-1}(\theta)$ and $h(\theta) = \mathrm{grad}\, g(\theta) = (g_1(\theta), \ldots, g_p(\theta))$, $g_i = \frac{\partial g}{\partial \theta_i}$; i.e.,

$$t \in U_g \Rightarrow \mathrm{Var}_\theta(t) \geq \mathrm{Var}_\theta(\tilde{t}) \geq \mathrm{Var}_\theta(t^*_{\theta,1}) = h(\theta) I^{-1}(\theta) h'(\theta),$$

where $\tilde{t}$ is the projection (of *any* $t \in U_g$) to W_θ and $t^*_{\theta,1}$ is the projection (again, of *any* $t \in U_g$) to $\mathrm{Span}\{1, dL/d\xi|_{\xi=0}\}$. Now (remembering that $\delta = \theta + \xi c^*$)

$$\frac{dL}{d\xi}\bigg|_{\xi=0} = \sum_{i=1}^p c_i^* L_i(\theta \mid s) =: L'$$

and, under P_θ (i.e., for $\xi = 0$) $1 \perp L'$, so $\{1, L'/||L'||\}$ is an orthonormal basis for $\mathrm{Span}\{1, L'\}$ and

$$t^*_{\theta,1} = g(\theta) \cdot 1 + \left(t, \frac{L'}{||L'||}\right) \cdot \frac{L'}{||L'||} = g(\theta) + \frac{1}{||L'||} \frac{dg}{d\xi}\bigg|_{\xi=0} \frac{L'}{||L'||}$$

$$= g(\theta) + \left(\sum_{i=1}^p c_i^* L_i(\theta \mid s)\right) \frac{\sum_i c_i^* g_i(\theta)}{\sum_{i,j} c_i^* I_{ij}(\theta) c_j^*} = g(\theta) + \left(\sum_{i=1}^p c_i^* L_i(\theta \mid s)\right) \frac{c^* h'}{c^* I c^{*'}}.$$

48

Note that $c^{*\prime} = I^{-1}h$, so $c^* I c^{*\prime} = h I^{-1} h' = c^* h'$ and so the above formula becomes

$$t_{\theta,1}^* = g(\theta) + \sum_{i=1}^{p} c_i^* L_i.$$

We have

$$\mathrm{Var}_\theta(t_{\theta,1}^*) = \frac{(\sum c_i^* g_i(\theta))^2}{(\sum_{i,j} c_i^* I_{ij}(\theta) c_j^*)} = \frac{(h I^{-1} h')^2}{(h I^{-1}) I (h I^{-1})'} = h I^{-1} h'.$$

More heuristic (as in the one-dimensional parameter case)

"ML estimates are nearly unbiased and nearly attain the bound in 12^p."

We assume that the ML estimate $\hat{\theta}$ of θ exists. Since Θ is open and $L(\cdot \mid s)$ is continuously differentiable, we have that

$$L_i(\hat{\theta}) = \left. \frac{\partial L(\theta \mid s)}{\partial \theta_i} \right|_{\theta = \hat{\theta}} = 0.$$

Choose and fix $\theta \in \Theta$, and regard it as the actual parameter value. If we assume that $\hat{\theta}$ is close to θ, then

$$L_i(\hat{\theta}) \approx L_i(\theta) + \sum_{j=1}^{p} (\hat{\theta}_j - \theta_j) L_{ji}(\theta), \quad i = 1, \ldots, p.$$

Assume that the sample is highly informative, i.e., that

$$L_{ji}(\theta \mid s) \approx -I_{ij}(\theta).$$

(We know that $E_\theta(L_{ji}(\theta \mid s)) = -I_{ji}(\theta)$. We are thus assuming that

$$\{L_{ji}\} = \{-I_{ji}(1 + \varepsilon_{ji})\},$$

where $\varepsilon_{ji}(\theta, s) \to 0$ in probability. This happens typically when the data is highly informative.) From this it follows that

$$L_i(\theta) \approx \sum_{j=1}^{p} (\hat{\theta}_j - \theta_j) I_{ji}(\theta), \quad i = 1, \ldots, p$$

– i.e., $(\hat{\theta} - \theta) I = (L_1, \ldots, L_p)$.

Definition. $L^{(1)}(\theta \mid s) := (L_1(\theta \mid s), \ldots, L_p(\theta \mid s))$ is the SCORE VECTOR.

Thus the ML estimate of a given g is

$$\hat{t}(s) = g(\hat{\theta}(s)) \approx g(\theta) + \sum_{j=1}^{p} (\hat{\theta}_j(s) - \theta_j) g_j(\theta) = g(\theta) + (\hat{\theta}(s) - \theta) h'(\theta)$$

$$\approx g(\theta) + L^{(1)}(\theta \mid s) I^{-1}(\theta) h'(\theta) = t_{\theta,1}^*$$

under P_θ. Since $E_\theta\big(L^{(1)}(\theta \mid s)\big) = 0$, we have $E_\theta(\hat{t}) \approx g(\theta)$. Since θ is arbitrary, $\hat{t}$ is approximately unbiased for g, i.e., $\hat{t} \dot\in U_g$. Since

$$\hat{t}(s) \approx g(\theta) + L^{(1)}(\theta \mid s)I^{-1}(\theta)h'(\theta) = g(\theta) + c^*\big(L^{(1)}(\theta \mid s)\big)'$$

under P_θ, we know that $\hat{t} \dot\in \mathrm{Span}\{1, L_1, \ldots, L_p\}$, so that $\hat{t} \approx t^*_{\theta,1}$ under P_θ and

$$\mathrm{Var}_\theta(\hat{t}) \approx \mathrm{Var}_\theta(t^*_{\theta,1}) = h(\theta)I^{-1}(\theta)h'(\theta).$$

This is, if true, remarkable, for it happens for *every* g and *every* $\theta \in \Theta$.

Example 3. Suppose that the X_i are iid $N(\mu, \sigma^2)$ and $\theta = (\theta_1, \theta_2) = (\mu, \sigma^2)$. Some functions g which may be of interest are $g(\theta) = \mu$, $g(\theta) = \sigma^2$ (or $g(\theta) = \sigma$), $g(\theta) = \mu/\sigma$ (or $g(\theta) = \sigma/\mu$, if $\mu \neq 0$) and $g(\theta) = $ the real number c such that $P_\theta(X_i < c) = \alpha$ (for some fixed $0 < \alpha < 1$) – i.e., $g(\theta) = \mu + z_\alpha\sigma$, where z_α is the normal α fractile.

Let us compute I. Since s consists of n iid parts, $I(\theta)$ for s is simply $nI_1(\theta)$, where $I_1(\theta)$ is I for X_1. If X_1 is the entire data, then

$$L = C - \frac{1}{2}\log\tau - \frac{1}{2\tau}(X_1 - \mu)^2,$$

where C is a constant and $\tau := \sigma^2 = \theta_2$; thus

$$L_1 = \frac{X_1 - \mu}{\tau} \quad \text{and} \quad L_2 = -\frac{1}{2\tau} + \frac{1}{2\tau^2}(X_1 - \mu)^2.$$

Homework 4

3. Check that

$$I_1(\theta) = \begin{pmatrix} 1/\tau & 0 \\ 0 & 1/2\tau^2 \end{pmatrix}.$$

Lecture 21

Example 3 (continued). We return to the situation $s = (X_1, \ldots, X_n)$; then

$$I(s) = n\begin{pmatrix} 1/\tau & 0 \\ 0 & 1/2\tau^2 \end{pmatrix} \quad \text{and} \quad I^{-1}(s) = \begin{pmatrix} \tau/n & 0 \\ 0 & 2\tau^2/n \end{pmatrix}.$$

Consider $g(\theta) = \mu = \theta_1$; then the most difficult one-dimensional problem is

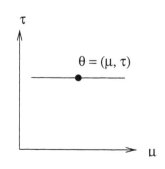

50

This one-dimensional problem is in a one-parameter exponential family with sufficient statistic $\overline{X}$, and $\overline{X}$ is a UMVUE in this one-dimensional problem which attains the C-R bound – i.e., $\overline{X}$ is unbiased and $\mathrm{Var}_\theta(\overline{X}) = h(\theta)I^{-1}(\theta)h'(\theta)$, where $h = (1,0)$; thus

$$\mathrm{Var}_\theta(\overline{X}) = \tau/n \ \forall \theta \in \Theta.$$

The following are some gs (and their corresponding C-R bounds) for which the C-R bound is *not* attained:

i. $g(\theta) = \sigma^2$; the C-R bound is $\frac{2\tau^2}{n}$.

ii. $g(\theta) = \sigma$; the C-R bound is $\frac{\tau}{2n}$.

iii. $g(\theta) = \mu + z_\alpha\sigma$, $h = (1, z_\alpha/2\sqrt{\tau})$; the C-R bound is $\frac{\tau}{n} + \tau\frac{z_\alpha^2}{2n}$.

To see this, it is enough to check case (i), since the reasoning for the other cases is similar. Here

$$\ell(\theta \mid s) = C\tau^{-n/2}e^{-\frac{1}{2\tau}[n(\overline{X}-\mu)^2 + nv]},$$

where C is a constant and $v = \frac{1}{n}\sum_{i=1}^n (X_i - \overline{X})^2$;

$$L(\theta \mid s) = C' - \frac{n}{2}\log\tau - \frac{1}{2\tau}\left[n(\overline{X} - \mu)^2 + nv\right],$$

where $C' = \log C$; $L_1(\theta \mid s) = \frac{n}{\tau}(\overline{X} - \mu)$ and

$$L_2(\theta \mid s) = -\frac{n}{2\tau} + \frac{1}{2\tau^2}\left[n(\overline{X} - \mu)^2 + nv\right].$$

Let $\delta = (\mu_*, \tau_*)$; then

$$E_\delta\big(L_1(\theta \mid s)\big) = \frac{n}{\tau}(\mu_* - \mu)$$

and

$$E_\delta\big(L_2(\theta \mid s)\big) = -\frac{n}{2\tau} + \frac{1}{2\tau^2}\left[\tau_*(n-1) + n\frac{\tau_*}{n} + n(\mu_* - \mu)^2\right]$$

$$= -\frac{n}{2\tau} + \frac{1}{2\tau^2}\left[n\tau_* + n(\mu_* - \mu)^2\right].$$

From these equations it is easily seen that there do *not* exist constants $a(\theta)$, $b(\theta)$ and $c(\theta)$ such that

$$E_\delta\big[a(\theta) + b(\theta)L_1(\theta \mid s) + c(\theta)L_2(\theta \mid s)\big] = \tau_*$$

for all $\delta = (\mu_*, \tau_*)$ – i.e., there is no unbiased estimate of τ_* in $\mathrm{Span}\{1, L_1(\theta \mid \cdot), L_2(\theta \mid \cdot)\}$, so that the C-R bound is not attainable for $g(\theta) = \tau$.

On the other hand, $\overline{X} = \mu + \frac{\tau}{n}L_1(\theta \mid s)$ is in $\mathrm{Span}\{1, L_1, L_2\}$ and is unbiased for μ, and so attains the C-R bound for μ. It is easy to check that the ML estimate is $\hat{\theta} = (\overline{X}, v)$, so the MLE for μ is $\overline{X}$; it is exactly unbiased, and its variation is the C-R bound. The MLE for $\tau = \sigma^2$ is $v = \frac{1}{n}\sum_{i=1}^{n}(X_i - \overline{X})^2$; we have that $E_\theta(v) = \frac{n-1}{n}\tau = \tau - \frac{\tau}{n}$ (note that $\frac{\tau}{n}$ is small when I is "large"),

$$\operatorname{Var}_\theta(v) = \frac{\tau^2}{n^2}\operatorname{Var}_\theta(X_{n-1}^2) = \frac{2(n-1)}{n^2}\tau^2,$$

which is *less* than the C-R bound $\frac{2\tau^2}{n}$ for τ (so v is *not* unbiased), and

$$\operatorname{MSE}_\theta(v) = \frac{2(n-1)}{n^2}\tau^2 + \frac{\tau^2}{n^2} = \frac{2\tau^2}{n} - \frac{\tau^2}{n^2} < \frac{2\tau^2}{n}.$$

Homework 4

4. The ML estimate for $\sigma = \sqrt{\tau}$ is $\sqrt{v}$. Show that $E_\theta(\sqrt{v}) = \sigma + o(1)$ and $\operatorname{Var}_\theta(\sqrt{v}) = \frac{\tau}{2n} + o(1)$ as $n \to \infty$. (HINT: z is an $X_k^2 \Leftrightarrow \frac{1}{2}z$ is a $\Gamma(k/2)$ variable. A $\Gamma(m)$ variable has density $\frac{e^{-x}x^{m-1}}{\Gamma(m)}$ in $(0, \infty)$. $\Gamma(m+1) = \sqrt{2\pi m}\cdot m^m e^{-m} + o(1/m)$ as $m \to \infty$, so

$$\frac{\Gamma(m+h)}{\Gamma(m)} = m^h(1 + o(1))$$

as $m \to \infty$ for a fixed h.)

Lecture 22

Note. In the general case of $(S, \mathcal{A}, P_\theta)$, $\theta \in \Theta$, the above considerations are somewhat more general than are required for strict unbiased estimation. In particular, associated with each $\theta \in \Theta$ there is a set W_θ of estimates which has the following properties:

Corollary to (8). If we are estimating a scalar $g(\theta)$ corresponding to any *estimate t, then there is an estimate $\tilde{t} \in W_\theta$ such that $E_\delta(t) = E_\delta(\tilde{t})$ for all $\delta \in \Theta$ and*

$$E_\theta\big(t - g(\theta)\big)^2 =: R_t(\theta) \geq R_{\tilde{t}}(\theta) := E_\theta\big(\tilde{t} - g(\theta)\big)^2,$$

with the inequality strict unless $P_\delta(t = \tilde{t}) = 1$ for all $\delta \in \Theta$.

In general, W_θ depends on θ and we must be content with $C = \bigcap_{\theta \in \Theta} W_\theta$. In some important special cases, however – for example, in an exponential family – W_θ is independent of θ. In any case, though, the MLE and related estimates have the property that, if "$I(\theta)$" is large, any smooth function $f(\hat{\theta})$ is approximately in W_θ for any fixed θ.

Example 3 (continued). $\theta = (\mu, \tau)$, where $\tau = \sigma^2$. Choose and fix θ; then what is W_θ? There are three methods available:

Method 1. Look at $\Omega_{\delta,\theta}$. W_θ is the subspace spanned by $\{\Omega_{\delta,\theta} : \delta \in \Theta\}$.

Method 2. (Let θ be real, under regularity conditions.) $\frac{d^j}{d\delta^j}\Omega_{\delta,\theta}\big|_{\delta=\theta} \in W_\theta$. This is the method which leads to the Cramér-Rao and Bhattacharya inequalities.

Method 3. (Due to Stein.) $\int_{\delta_1}^{\delta_2} \Omega_{\delta,\theta} d\delta \in W_\theta$.

We use Method 2. Since $\ell(\theta \mid s) = e^{L(\theta\mid s)}$, we have $\ell_i(\theta \mid s) = e^{L(\theta\mid s)} L_i(\theta \mid s)$,

$$\ell_{ij}(\theta \mid s) = e^{L(\theta\mid s)} \big[L_{ij}(\theta \mid s) + L_i(\theta \mid s) L_j(\theta \mid s) \big],$$

etc., and hence $\ell_i/\ell = L_i$, $\ell_{ij}/\ell = L_{ij} + L_i L_j$, etc. Thus ℓ_i/ℓ, ℓ_{ij}/ℓ, etc. are in W_θ. Here we have

$$L_1 = \frac{n(\overline{X} - \mu)}{\tau} \qquad\qquad L_2 = \frac{n[v + (\overline{X} - \mu)^2]}{2\tau^2} - \frac{n}{2\tau}$$

$$L_{11} = -\frac{n}{\tau} \qquad\qquad L_{21} = -\frac{n(\overline{X} - \mu)}{\tau^2}$$

$$L_{12} = -\frac{n(\overline{X} - \mu)}{\tau^2} \qquad L_{22} = -\frac{n[v + (\overline{X} - \mu)^2]}{\tau^3} - \frac{n}{2\tau^2}.$$

Since $\ell_{11}/\ell = L_{11} + L_1^2$ is an affine function of $(\overline{X} - \mu)^2$, we have

$$\mathrm{Span}\{1, \overline{X}, v, (\overline{X} - \mu)^2\} = \mathrm{Span}\{1, L_1(\theta \mid \cdot), L_2(\theta \mid \cdot), \ell_{11}(\theta \mid \cdot)/\ell(\theta \mid \cdot)\} \subseteq W_\theta,$$

whence $\overline{X}$ is the LMVUE of $E_\delta(\overline{X}) = \mu_*$, v is the LMVUE of $E_\delta(v) = \frac{n-1}{n}\tau_*$ and $\frac{nv}{n-1}$ is the LMVUE of $E_\delta\big(nv/(n-1)\big) = \tau_*$ (remember $\delta = (\mu_*, \tau_*)$.) Since $\overline{X}$, v and $\frac{nv}{n-1}$ do not depend on θ, they are in fact in $C = \bigcap_{\theta \in \Theta} W_\theta$ and hence are the UMVUEs of their expected values. (Neither $\sqrt{v}$ nor $\frac{\overline{X}}{\sqrt{v}}$ (the latter is the MLE of μ/σ) is available by this method, but one can show by the above method that any function of $\overline{X}$ and v is in C. If Θ is the set of all pairs (μ, σ^2), then we are in the two-parameter exponential family case and a result to be stated later applies.)

Regularity conditions

Θ is open in $\mathbb{R}^p$ and $dP_\theta(s) = \ell(\theta \mid s) d\mu(s)$.

Condition 1^p. For each s, $\ell(\cdot \mid s)$ is a positive continuously differentiable function of θ.

Condition 2^p. Given any $\theta \in \Theta$, we may find an $\varepsilon = \varepsilon(\theta) > 0$ such that

$$\max\{|L_j(\delta \mid s)| : |\delta_i - \theta_i| \le \varepsilon\} \in V_\theta$$

(i.e., the function is square-integrable with respect to P_θ), or at least

$$\frac{\max\{|\ell_j(\delta \mid s)| : |\delta_i - \theta_i| \le \varepsilon\}}{\ell(\theta \mid s)} \in V_\theta.$$

Let $I(\theta) = E_\theta\big(L_i(\theta \mid s) L_j(\theta \mid s)\big)$.

Condition 3^p. For each θ, $I(\theta)$ is positive definite.

12^pE. a. For each θ, $1, L_1(\theta \mid s), \ldots, L_p(\theta \mid s) \subseteq W_\theta$, and $1 \perp L_j(\theta \mid s)$ in V_θ for $j = 1, \ldots, p$.

b. If U_g is non-empty, then g is differentiable and the projection of any $t \in U_g$ to $\mathrm{Span}\{1, L_1, \ldots, L_p\}$ (which is the projection of $\tilde{t}$ to $\mathrm{Span}\{1, L_1, \ldots, L_p\}$) is

$$t^*_{\theta,1} = g(\theta) + h(\theta)I^{-1}(\theta)\big(L_1(\theta \mid s), \ldots, L_p(\theta \mid s)\big)',$$

where $h(\theta) = \mathrm{grad}\, g(\theta)$.

c. If $t \in U_g$, then $\mathrm{Var}_\theta(t) \geq h(\theta)I^{-1}(\theta)h'(\theta)$ for all $\theta \in \Theta$.

Proof. The proof is left as an exercise for the reader. See the proof in the case $p = 1$ and use Approach 1 rather than Approach 2.

Note also that $g(\theta)$ is a projection of t to $\mathrm{Span}\{1\}$ and that 1 is orthogonal to $L_1, \ldots, L_p$, so that the projection of $t - g(\theta)$ to $\mathrm{Span}\{1, L_1, \ldots, L_p\}$ is the same as its projection to $\mathrm{Span}\{L_1, \ldots, L_p\}$. Thus

$$\mathrm{Var}_\theta\big(t - g(\theta)\big) \geq E_\theta\big(\text{projection of } t - g(\theta) \text{ to } \mathrm{Span}\{L_1, \ldots, L_p\}\big)^2$$
$$= E_\theta\big(hI^{-1}(L_1, \ldots, L_p)'[hI^{-1}(L_1, \ldots, L_p)']'\big)$$
$$= E_\theta\big(hI^{-1}(L_1, \ldots, L_p)'(L_1, \ldots, L_p)I^{-1}h'\big) = hI^{-1}h'.$$

Lecture 23

Note. In the case when Θ is open in $\mathbb{R}^p$, $g : \Theta \to \mathbb{R}'$ is differentiable and conditions 1^p–3^p are satisfied, then, for any estimate t,

$$R_t(\theta) := E_\theta\big(t(s) - g(\theta)\big)^2 \geq \beta_t(\theta)I^{-1}(\theta)\beta'_t(\theta) + \big[b_t(\theta)\big]^2,$$

where $b_t(\theta) := E_\theta(t) - g(\theta)$ and $\beta_t(\theta) := \mathrm{grad}\, E_\theta(t) = \mathrm{grad}\, g(\theta) + \mathrm{grad}\, b_t(\theta)$.

Proof. Let $\gamma(\delta) = E_\delta(t)$, so that $t \in U_\gamma$. Then

$$R_t(\theta) = \mathrm{Var}_\theta(t) + \big[b_t(\theta)\big]^2 \geq \big[\mathrm{grad}\, \gamma(\theta)\big]I^{-1}(\theta)\big[\mathrm{grad}\, \gamma(\theta)\big]'$$

by C-R bound. □

This result is useful even in case $p = 1$ – see, for example, the proof of the admissibility of $\hat{\theta}$ in Example 1(a) in Lehmann (1983, *Theory of point estimation*).

54

On the distance between θ and δ

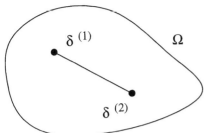

Should one use the Euclidean distance d_1? What is really of interest is the "distance" between $P_{\delta^{(1)}}$ and $P_{\delta^{(2)}}$ – given, say, by

$$d_2(\delta^{(1)}, \delta^{(2)}) = \sup_{A \in \mathcal{A}} |P_{\delta^{(1)}}(A) - P_{\delta^{(2)}}(A)| = \frac{1}{2} \int_S |\ell(\delta^{(1)} \mid s) - \ell(\delta^{(2)} \mid s)| ds$$

or

$$d_3(\delta^{(1)}, \delta^{(2)}) = \int_S \left(\sqrt{\ell(\delta^{(1)} \mid s)} - \sqrt{\ell(\delta^{(2)} \mid s)} \right)^2 d\mu(s).$$

The distance d_3 is used in E. J. G. Pitman (1979, *Some basic theory of statistic inference*). It is related to the Fisher information in the following way:

Suppose that we want to distiguish between $P_{\delta^{(1)}}$ and $P_{\delta^{(2)}}$ on the basis of s. Instead of a hypothesis-testing approach, let us choose a real-valued function $t(s)$. What is the difference between $\delta^{(1)}$ and $\delta^{(2)}$ on the basis of t?

Regard t as an estimate of $g(\delta) := E_\delta(t)$. Then $|g(\delta^{(1)}) - g(\delta^{(2)})|$ might be taken as a measure of the distance between $\delta^{(1)}$ and $\delta^{(2)}$ on the basis of t. It is, however, more plausible to use the standardized versions

$$\frac{1}{\mathrm{SD}_{\delta^{(1)}}(t)} |g(\delta^{(1)}) - g(\delta^{(2)})| \quad \text{and} \quad \frac{1}{\mathrm{SD}_{\delta^{(2)}}(t)} |g(\delta^{(1)}) - g(\delta^{(2)})|,$$

especially if t is approximately normally distributed.

Now choose and fix $\theta \in \Theta$ and restrict δ to a small neighborhood of θ. Then $\mathrm{Var}_\delta(t) \approx \mathrm{Var}_\theta(t)$, and hence the distance (between $\delta^{(1)}$ and $\delta^{(2)}$, on the basis of t) is approximately

$$\frac{|g(\delta^{(1)}) - g(\delta^{(2)})|}{\sqrt{\mathrm{Var}_\theta(t)}} =: d_{t,\theta}(\delta^{(1)}, \delta^{(2)}).$$

Since the distance should be "intrinsic", we should maximize it with respect to t. First, we maximize $d_{t,\theta}$ with respect to t with the expectation function g fixed to get

$$\frac{|g(\delta^{(1)}) - g(\delta^{(2)})|}{\sqrt{(\mathrm{grad}\, g(\theta)) I^{-1}(\theta) (\mathrm{grad}\, g(\theta))'}}.$$

With $\delta^{(1)} \to \theta$ and $\delta^{(2)} \to \theta$, this is approximately

$$\frac{|(\delta^{(1)} - \delta^{(2)})[\mathrm{grad}\, g(\theta)]'|}{\sqrt{(\mathrm{grad}\, g(\theta)) I^{-1}(\theta) (\mathrm{grad}\, g(\theta))'}}.$$

Next, maximize the square of this with respect to $h(\theta) = \operatorname{grad} g(\theta)$, which then leads to the squared distance

$$D_\theta^2(\delta^{(1)}, \delta^{(2)}) = (\delta^{(2)} - \delta^{(1)}) I(\theta) (\delta^{(2)} - \delta^{(1)})'.$$

The distance D_θ is called the LOCAL FISHER METRIC in the vicinity of θ. It is the distance between $P_{\delta^{(1)}}$ and $P_{\delta^{(2)}}$ as measured in standard units for a real-valued statistic of the form $g(\hat{\theta})$, where g is suitably chosen so that $\operatorname{grad} g(\theta) = (\delta^{(2)} - \delta^{(1)}) I(\theta)$.

Example 1(a). Let $n = 1$, $s \sim N(\theta, \sigma^2)$ and $\theta \in \Theta = (-\infty, \infty)$, where σ^2 is a fixed known quantity. Then $I(\theta) = 1/\sigma^2$ for all θ,

$$D_\theta^2(\delta^{(2)}, \delta^{(1)}) = \frac{(\delta^{(2)} - \delta^{(1)})^2}{\sigma^2}$$

and

$$D = \frac{|\delta^{(2)} - \delta^{(1)}|}{\sigma} = \left| \frac{\text{mean of } P_{\delta^{(1)}} - \text{mean of } P_{\delta^{(2)}}}{\text{common SD}} \right|.$$

If $n > 1$ and $s = (X_1, \ldots, X_n)$ with the X_i iid, then

$$D_\theta(\delta^{(1)}, \delta^{(2)}) = \sqrt{n} \left| \frac{\delta^{(2)} - \delta^{(1)}}{\sigma} \right|.$$

For fixed $\theta \in \mathbb{R}^p$, D_θ is the metric derived from the inner product

$$(u|v)_* := \sum_{i,j} u_i I_{ij}(\theta) v_j = u I(\theta) v',$$

which has been used before. *Exercise* (informal): Look at D_θ in Example 3, $N(\theta_1, \theta_2)$.
Example 4. $Y \in \mathbb{R}^k$ has the $N_k(\theta, \Sigma)$ distribution and density

$$\ell(\theta \mid y) = \frac{1}{(2\pi)^{k/2} |\Sigma|^{k/2}} e^{-\frac{1}{2}(y-\theta)\Sigma^{-1}(y-\theta)'}$$

with respect to Lebesgue measure. With this density, θ and Σ are respectively the mean and covariance matrices of Y. Show that $I(\theta) = \Sigma^{-1}$ for all θ and hence $D_\theta^2(\delta^{(2)}, \delta^{(1)})$ is the fixed square distance $(\delta^{(2)} - \delta^{(1)}) \Sigma^{-1} (\delta^{(2)} - \delta^{(1)})$.

Lecture 24

Note. A sufficient condition for 13^p – i.e., the equality $I(\theta) = -\{E_\theta(L_{ij}(\theta \mid s))\}$ – is that, given any $\theta \in \Theta$, we may find an $\varepsilon = \varepsilon(\theta) > 0$ such that

$$\max\{|\ell_{ij}(\delta \mid s)/\ell(\delta \mid s)| : |\delta_i - \theta_i| \leq \varepsilon\},$$

or at least

$$\max\{|\ell_{ij}(\delta \mid s)| : |\delta_i - \theta_i| \leq \varepsilon\}/\ell(\theta \mid s),$$

be P_θ integrable (for $i, j = 1, \ldots, p$).

Note. The theory extends to estimation of vector-valued functions – for example, if $u(s) = (u_1(s), \ldots, u_p(s))$ is an unbiased estimate of θ and $\mathrm{Var}_\theta(u_i) < +\infty$ for each $i = 1, \ldots, p$ and $\theta \in \Theta$, then $\mathrm{Cov}_\theta(u) - I^{-1}(\theta)$ is positive semidefinite for each $\theta \in \Theta$.

Proof. Fix $a = (a_1, \ldots, a_p) \in \mathbb{R}^p$ and define $g(\theta) = \sum_{i=1}^p a_i \theta_i = a\theta'$. Then $t(s) = au'(s)$ is an unbiased estimate of g. Since $\mathrm{grad}\, g(\theta) = a$, we have

$$\mathrm{Var}_\theta(t) = a\,\mathrm{Cov}_\theta(u)a' \geq aI^{-1}(\theta)a',$$

so that ($a \in \mathbb{R}^p$ having been arbitrary) $\mathrm{Cov}_\theta(u) - I^{-1}(\theta)$ is positive semidefinite. □

Definition. $(S, \mathcal{A}, P_\theta)$, $\theta \in \Theta \subseteq \mathbb{R}^p$ is a (p-parameter) EXPONENTIAL FAMILY with statistic $T = (T_1, \ldots, T_p) : S \to \mathbb{R}^p$ if $dP_\theta(s) = \ell(\theta \mid s)d\mu(s)$, where

$$\ell(\theta \mid s) = C(s)e^{B_1(\theta)T_1(s) + \cdots + B_p(\theta)T_p(s) + A(\theta)}.$$

The family is NON-DEGENERATE at a particular $\theta \in \Theta$ if

$$\left\{ (B_1(\delta) - B_1(\theta), \ldots, B_p(\delta) - B_p(\theta)) : \delta \in \Theta \right\}$$

contains a neighborhood of $0 = (0, \ldots, 0)$.

We assume non-degeneracy at each $\theta \in \Theta$.

Exercise: Check that Example 1(a) is a non-degenerate exponential family with $p = 1$, with $T_1 = \overline{X}$ if $\Theta = \mathbb{R}^1$; Example 2(a) is a non-degenerate exponential family with $p = 1$, $T_1 = \overline{X}$ and $\Theta = (0, 1)$; Example 2(b) is a non-degenerate exponential family with $p = 1$, $T_1 = N$ and $\Theta = (0, 1)$; Example 3 is a two-parameter non-degenerate exponential family with $T_1 = \sum X_i$, $T_2 = \sum X_i^2$ and

$$\Theta = \{(\mu, \tau) : -\infty < \mu < +\infty \text{ and } 0 < \tau < +\infty\};$$

and Example 4 is a k-parameter exponential family with $T = \sum y_i = (T_1, \ldots, T_k)$.

15^p.   a. For each $\theta \in \Theta$, W_θ is the space of all Borel functions of $T = (T_1, \ldots, T_p)$ which are in V_θ.

   b. $C = \bigcap_{\theta \in \Theta} W_\theta$ is the class of all UMVUE – i.e., the class of all Borel functions of T which are in $L^2(P_\theta)$ for all $\theta \in \Theta$.

   c. For any g such that U_g is non-empty, there exists an essentially unique estimate $\tilde{t} = \tilde{t}(T) \in C \cap U_g$.

   d. $\tilde{t} = E_\theta(t \mid T)$ for all $t \in U_g$ and $\theta \in \Theta$.

   e. For all $A \subseteq S$, $E_\theta(I_A \mid T) = P_\theta(A \mid T)$ (essentially) is the same for each $\theta \in \Theta$, i.e., T is a sufficient statistic.

   f. T is a complete statistic.

Proof.

a. Choose $\theta \in \Theta$ and write $\xi_i = B_i(\delta) - B_i(\theta)$. Then

$$\Omega_{\delta,\theta} = e^{\sum_{i=1}^{p} \xi_i T_i(s) - K_\theta(\xi_1, \ldots, \xi_p)},$$

where $K_\theta(\xi_1, \ldots, \xi_p) = \log E_\theta(e^{\sum \xi_i T_i(s)})$ is the cumulant generating function of T at $(\xi_1, \ldots, \xi_p)$ under P_θ. Non-degeneracy means that

$$K_\theta(\xi_1, \ldots, \xi_p) < +\infty$$

for $(\xi_1, \ldots, \xi_p)$ in a neighborhood of 0, and hence W_θ contains all functions $e^{\sum \xi_i T_i}$ for $(\xi_1, \ldots, \xi_p)$ in a neighborhood of 0. By differentiation, we find that W_θ contains all polynomials in $T_1, \ldots, T_p$, so W_θ contains all Borel functions of T which belong to V_θ.

On the other hand, since each $\Omega_{\delta,\theta}$ is a Borel function of T, every function in W_θ is such; so (a) is proved.

b. This follows from (a) and (9).

c. This follows from (a) and (8).

d. This follows from (a) and (8) and the fact that, if W is the space of all functions of T, projection to W is the conditional expectation given T.

e. This follows from (c) and (d) by letting $g(\theta) = P_\theta(A)$.

f. Suppose $E_\theta h = 0$ and $E_\theta h^2 < +\infty$ for all $\theta \in \Theta$. Then $h(T)$ is the UMVUE of $g(\theta) = 0$; but 0 is an unbiased estimate of this g, so $\text{Var}_\theta h = 0$ for all $\theta \in \Theta$ and hence $P_\theta(h = 0) = 1$ for all $\theta \in \Theta$. □

Chapter 7

Lecture 25

Using the score function (or vector)

Assume the usual setting, $(S, \mathcal{A}, P_\theta)$, $\theta \in \Theta \subseteq \mathbb{R}^p$.

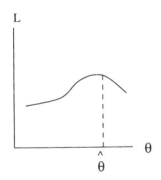

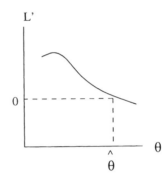

First consider the case $p = 1$. Let $u(s)$ be a trial solution of $L'(\theta \mid s) = 0$. Assume that $\hat\theta - \theta = O(1/\sqrt{I(\theta)})$ and I is large. (Here θ is the true parameter, $E_\theta(\hat\theta) \approx 0$ and $\text{Var}_\theta(\hat\theta) \approx 1/I(\theta)$.) Assume that u is not very inaccurate in the sense that, for any θ, $u(s) - \theta = O(1/\sqrt{I(\theta)})$. Then $\hat\theta - u = O(1/\sqrt{I(\theta)})$ under θ,

$$0 = L'\big(\hat\theta(s) \mid s\big) = L'(u(s) \mid s) + \big(\hat\theta(s) - u(s)\big)L''(u(s) \mid s) + O(1/I(\theta))$$

and

$$\hat\theta(s) = u(s) + \left(-\frac{1}{L''(u(s) \mid s)}\right)L'(u(s) \mid s) + O(1/I(\theta)).$$

Dropping the last term (order $1/I(\theta)$), we obtain the 'first Newton iterate' for solving $L'(\theta \mid s) = 0$.

Application 1. Let $u^{(0)}(s)$ be a trial solution of $L'(\theta \mid s) = 0$. Let

$$u^{(j+1)}(s) = u^{(j)}(s) + \left(-\frac{1}{L''(u^{(j)}(s) \mid s)}\right)L'(u^{(j)}(s) \mid s).$$

One hopes that $u^{(j)}(s) \to \hat\theta(s)$.

A variant of this approach consists in taking

$$u^{(j+1)}(s) = u^{(j)}(s) + \frac{1}{I(u^{(j)}(s))} L'(u^{(j)}(s) \mid s)$$

(since typically $-L''(\theta \mid s)/I(\theta) \approx 1$ if $I(\theta)$ is large).

Suppose we do not think it worthwhile to find $\hat{\theta}$ exactly.

Application 2. Start with a plausible estimate $u(s)$ of θ, and improve it to

$$u^*(s) = u(s) + \left(-\frac{1}{L''(u(s) \mid s)} \right) L'(u(s) \mid s)$$

or

$$u^{**}(s) = u(s) + \frac{1}{I(\theta)} L'(u(s) \mid s).$$

If $u - \theta = O(1/\sqrt{I(\theta)})$ and $E_\theta(u) - \theta = O(1/\sqrt{I(\theta)})$, then the first iterates have the same properties as $\hat{\theta}$, i.e., $u^* - \theta$ and $u^{**} - \theta$ are of order $1/\sqrt{I(\theta)}$ and $\mathrm{Var}_\theta(u^*)$ and $\mathrm{Var}_\theta(i^{**})$ are $b_1(\theta) = 1/I(\theta)$.

The case $p \geq 1$

Let $u(s) = (u_1(s), \ldots, u_p(s)) : S \to \Theta \subseteq \mathbb{R}^p$ be some plausible estimate of θ. Then

$$u^*(s) = u(s) + \left\{ -L_{ij}(u(s) \mid s) \right\}^{-1} \left\{ \mathrm{grad}\, L(\theta \mid s) \big|_{\theta=u(s)} \right\}$$

and

$$u^{**}(s) = u(s) + I^{-1}(u(s)) \left\{ \mathrm{grad}\, L(\theta \mid s) \big|_{\theta=u(s)} \right\}$$

are versions of the first iteration of the Newton-Raphson method for solving $\mathrm{grad}\, L(\theta \mid s) = 0$.

Let $||\cdot||$ be the Euclidean norm. If $||u - \theta||$ and $||\hat{\theta} - \theta||$ are of the same order and $E_\theta(\hat{\theta}) \approx \theta$ and $\mathrm{Cov}_\theta(\hat{\theta}(s)) \approx I^{-1}(\theta)$, then u^* and u^{**} also have these properties – i.e., $E_\theta(u^*) \approx \theta$ and $\mathrm{Cov}_\theta(u^*(s)) \approx I^{-1}(\theta)$ (and similarly for u^{**}).

Example 1. $s = (X_1, \ldots, X_n)$, with the X_i iid with density $f(x - \theta)$ for $\theta \in \mathbb{R}^1$.

a. f is the normal density. $I(\theta) = n$, $L'(\theta \mid s) = n(\overline{X} - \theta)$ and $L''(\theta \mid s) = -n$. For any u, the first iteration gives $u^* = \overline{X} = u^{**}$.

b. $f(x) = \frac{1}{2}e^{-|x|}$. We know from the homework that $\hat{\theta}$ is the median of $X_1, \ldots, X_n$. Here L' and I do not exist, but the Chapman-Robbins bound gives $\mathrm{Var}_\theta(t) \geq \frac{1}{n}$ for any unbiased estimate t of g. Show that $\mathrm{Var}_\theta(\hat{\theta}) = \frac{1}{n} + O(\frac{1}{n^2})$. (Note that

$$\mathrm{Var}_\theta(\overline{X}) = \frac{1}{n}\mathrm{Var}_\theta(X_1) = \frac{1}{n}\int \frac{x^2}{2}e^{-|x|}dx = \frac{1}{n}\int_0^\infty x^2 e^{-x}dx = \frac{\Gamma(3)}{n} = \frac{2}{n},$$

so that the variance bound is true for $\overline{X}$.)

c. $f(x) = \frac{1}{\pi}\frac{1}{1+x^2}$. Here $I_1(\theta) = \frac{1}{2}$ and $I(\theta) = \frac{n}{2}$. $\hat{\theta}$ is hard to find (there are many roots of $L'(\theta \mid s) = 0$).

$$L(\theta \mid s) = C - \sum_{i=1}^{n} \log\left[1 + (X_i - \theta)^2\right],$$

where C is a constant, and

$$L'(\theta \mid s) = \sum_{i=1}^{n} \frac{2(X_i - \theta)}{1 + (X_i - \theta)^2}.$$

Let $u(s)$ be the median of $\{X_1, \ldots, X_n\}$; then

$$u^{**}(s) = u(s) + \frac{4}{n}\sum_{i=1}^{n} \frac{X_i - u(s)}{1 + (X_i - u(s))^2}.$$

Since it is true that $u(s) - \theta$ is $O(1/\sqrt{n})$, we have $E_\theta(u^{**}) \approx \theta$ and $\mathrm{Var}_\theta(u^{**}) \approx \frac{2}{n}$, the information bound.

e. $f(x) = ae^{-bx^4}$, $a, b > 0$, and $\mathrm{Var}(x) = 1$. Here, as in (c) above, it is difficult to find W_θ, and $W_{\theta,1}$ and $W_{\theta,2}$ look awful. $\overline{X}$ is a plausible estimate since $E_\theta(\overline{X}) = \theta$ and $\mathrm{Var}_\theta(\overline{X}) = \frac{1}{n} = O(1/I(\theta))$ $(I(\theta) = n)$.

The most important differences among the above four densities are the different tail behaviors:

1(e). SHORT TAIL: Here a good estimate gives more weight to the extreme values than to the central values.

1(a). NORMAL: Here the best estimate $\overline{X}$ gives equal weight to all observations.

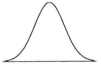

1(b). DOUBLE EXPONENTIAL: Here the best estimate, the median, gives weights concentrated in the middle.

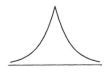

1(c). CAUCHY: Here the optimal estimate(s) is (are) unknown.

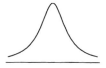

Lecture 26

Continuing Example 1(e)

$$\ell(\theta \mid s) = a^n e^{-b\sum_{i=1}^{n}(X_i-\theta)^4} = \varphi(s)e^{-b[-4\theta\sum X_i^3 + 6\theta^2\sum X_i^2 - 4\theta^3\sum X_i] + A(\theta)}$$

$$= \varphi(s)e^{B_1(\theta)m_3' + B_2(\theta)m_2' + B_1(\theta)m_1' + A(\theta)},$$

where $m_j' = \frac{1}{n}\sum_{i=1}^{n}X_i^j$ for $j = 1, 2, 3$ (notice that $m_1' = \overline{X}$). This is not a three-parameter exponential family but a curved exponential family; but (m_1', m_2', m_3') is equivalent to $(\overline{X}, m_2, m_3)$, where $m_j = \frac{1}{n}\sum_{i=1}^{n}(X_i - \overline{X})^j$ for $j = 2, 3$, which is the minimal sufficient statistic – i.e., $(\overline{X}, m_2, m_3)$ is an adequate summary of data (for any statistical purpose) and nothing less will do. (In Example 1(a), $\overline{X}$ is the minimal sufficient statistic, and, in Example 3, $(\overline{X}, m_2)$ is the minimal sufficient statistic.)
$L'(\theta) = 4b\sum_{i=1}^{n}(X_i - \theta)^3$. Let $\hat{\theta} = \overline{X} + zm_2^{1/2}$. Since $L'(\hat{\theta}) = 0$, we have

$$z + \frac{1}{3}z^3 = \frac{1}{3}\gamma_1,$$

where $\gamma_1 = m_3/m_2^{3/2}$ is the sample coefficient of kurtosis.

There are several approaches to getting $\hat{\theta}$:

Approach 3. Get an explicit form of z from the equation in z above, and substitute it into the expression for $\hat{\theta}$ in terms of z.

Approach 4. The graphic method:

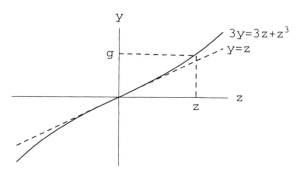

(In the picture above, $g = \gamma_1/3$.) $0 < z < \frac{1}{3}\gamma_1$ if $\gamma_1 > 0$ and $\frac{1}{3}\gamma_1 < z < 0$ if $\gamma_1 < 0$; so a solution is

$$z = \frac{1}{3}\gamma_1 - \frac{\eta}{27}\gamma_1^3,$$

where $0 < \eta < 1$.

62

Approach 5.

$$\hat{\theta} \approx \overline{X} + \frac{1}{3}\gamma_1 m_2^{1/2} = \overline{X} + \frac{1}{3}\frac{m_3}{m_2}.$$

Note that, if n is large, then $m_2 \approx 1$ (since $\operatorname{Var}_\theta(X_1) = 1$) and so $\hat{\theta} \approx \overline{X} + \frac{1}{3}m_3$ (so outliers are given more weight than given by $\overline{X}$). Here $I(\theta) = n$ and $\operatorname{Var}_\theta(\overline{X}) = \frac{1}{n} = O\left(\frac{1}{I(\theta)}\right)$, so $\overline{X}$ is an acceptable starting value for approximating $\hat{\theta}$.

Approach 6. $u^* = \overline{X} + \frac{1}{3}\frac{m_3}{m_2}$ and $u^{**} = \overline{X} + \frac{1}{3}m_3$ (please check). It is not easy to find the exact properties of $\hat{\theta}$, u^* and u^{**}, but u^{**} is the easiest to examine.

Homework 5

1. Show that $E_\theta(u^{**}) = \theta$ and

$$\operatorname{Var}_\theta(u^{**}) = b_1(\theta) + O\left(\frac{1}{n^2}\right) = \frac{1}{12bn} + O\left(\frac{1}{n^2}\right) = \frac{1}{1.37n} + O\left(\frac{1}{n^2}\right)$$

(so that $E_\theta(m^3) = 0$ and $\operatorname{Cov}_\theta(\overline{X}, m_3) < 0$).

Since m_3 is a function of the (minimal) sufficient statistic $T(s) = (\overline{X}, m_2, m_3)$, this statistic is not complete. Since $\operatorname{Cov}_\theta(\overline{X}, m_3) \neq 0$ (m_3 is an unbiased estimate of 0), we know that $\overline{X}$ is not even locally MVUE. (See Kendall and Stuart, vol. I, for "standard error of moments". A good reference to the use of the score function in general is C. R. Rao's *Linear Statistical Inference.*)

Example 5. Our state space is $\{1, 2\}$ and the transition probability matrix is

$$\begin{pmatrix} \theta_{11} & \theta_{12} \\ \theta_{21} & \theta_{22} \end{pmatrix} = \begin{pmatrix} \theta_1 & 1 - \theta_1 \\ 1 - \theta_2 & \theta_2 \end{pmatrix}.$$

Suppose first that $\Theta = (0,1) \times (0,1)$ and that a Markoff chain with transition probability matrix as above starts at '1' and is observed for n one-step transitions. Thus $s = (X_0, X_1, \ldots, X_n)$, where $X_0 \equiv 1$, and

$$\ell(\theta \mid s) = \prod_{i,j=1,2} \theta_{ij}^{f_{ij}(s)} = \theta_1^{f_{11}(s)}(1 - \theta_1)^{f_{12}(s)} \theta_2^{f_{22}(s)}(1 - \theta_2)^{f_{21}(s)},$$

where $f_{ij}(s)$ is the number of one-step transitions from i to j in s. Since $f_{11} + f_{12} + f_{22} + f_{21} = n$, we have a three-dimensional minimal sufficient statistic and two parameters. If $f_{21} + f_{22} > 0$, $f_{11} > 0$ and $f_{22} > 0$, then we have (noticing that $f_{11} + f_{12} > 0$) $\hat{\theta} = (\hat{\theta}_1, \hat{\theta}_2)$, where $\hat{\theta}_1 = \frac{f_{11}}{f_{11}+f_{12}}$ and $\hat{\theta}_2 = \frac{f_{22}}{f_{21}+f_{22}}$. Since

$$L_1 = \frac{f_{11}}{\theta_1} - \frac{f_{12}}{1 - \theta_1}, \quad L_2 = \frac{f_{22}}{\theta_2} - \frac{f_{21}}{1 - \theta_2}, \quad L_{11} = -\frac{f_{11}}{\theta_1^2} - \frac{f_{12}}{(1 - \theta_1)^2}, \ldots,$$

we have $E_\theta(L_1) = 0 = E_\theta(L_2)$ (since $(1 - \theta_1)E_\theta(f_{11}) = \theta_1 E_\theta(f_{12})$, etc.) and

$$I(\theta) = \begin{pmatrix} E_\theta(f_{11}/\theta_1^2 + f_{12}/(1 - \theta_1)^2) & 0 \\ 0 & E_\theta(f_{21}/(1 - \theta_2)^2 + f_{22}/\theta_2^2) \end{pmatrix}.$$

It is known that
$$E_\theta(f_{ij}) = n\pi_i(\theta)\theta_{ij} + o(n) \quad \text{as } n \to \infty$$
where $\pi_1(\theta)$ and $\pi_2(\theta)$ are the stationary distrtibution over $\{1,2\}$ and

$$\pi_1(\theta) = \frac{1-\theta_2}{2-(\theta_1+\theta_2)} \quad \text{and} \quad \pi_2(\theta) = \frac{1-\theta_1}{2-(\theta_1+\theta_2)},$$

so

$$I(\theta) = n \begin{pmatrix} \pi_1(\theta)/\theta_1(1-\theta_1) & 0 \\ 0 & \pi_2(\theta)/\theta_2(1-\theta_2) \end{pmatrix} + o(n).$$

The information bound for the variances of estimates of θ_1 is $\frac{\theta_1(1-\theta_1)}{n\pi_1(\theta)}$ (and similarly for θ_2). Is $\mathrm{Var}_\theta(\hat{\theta}_1) \approx \frac{\theta_1(1-\theta_1)}{n\pi_1(\theta)}$? It can be shown (though not easily) that $\mathrm{Var}_\theta(\hat{\theta}_1) = b_1(\theta) + o(1/n)$ as $n \to \infty$, where $b_1(\theta)$ is the C-R bound.

Lecture 27

In Example 5, Θ is an open unit square consisting of points (θ_1, θ_2). Let $\hat{\theta}_1 = \frac{f_{11}}{f_{11}+f_{12}}$ and $\hat{\theta}_2 = \frac{f_{22}}{f_{21}+f_{22}}$ if $f_{ij} > 0$ for all i, j. Otherwise, let $\hat{\theta}_2$ be arbitrary – say $\frac{1}{2}$, for convenience. It can be shown that

$$P_\theta(f_{ij} > 0 \; \forall i, j) \geq 1 - \left[p(\theta)\right]^n$$

for all sufficiently large n and some fixed $0 < p(\theta) < 1$. Hence we can ignore the case $f_{ij} = 0$ in the computation of $E_\theta(\hat{\theta})$ and $\mathrm{Var}_\theta(\hat{\theta})$.

Suppose we know that $\theta_2 = k\theta_1$ for some $0 < k < \infty$; then now $\Theta = \{\theta_1 : 0 < \theta_1 < 1/k\}$ and

$$L \propto f_{11} \log\theta_1 + f_{12}\log(1-\theta_1) + f_{21}\log(1-k\theta_1) + f_{22}\log k\theta_1.$$

Exercise: Show that I in the present case is greater than I in the previous case, for sufficiently large n. (Recall that $E_\theta(f_{ij}) = n\pi_i(\theta)\theta_{ij} + o(n)$.)

The equation for $\hat{\theta}_1$ is now a cubic. We can solve it explicitly, or we can approximate it by $v = u^*$ or u^{**}, with $u = \frac{f_{11}}{f_{11}+f_{12}}$ (say). Then we have $E_\theta(v) = \theta_1 + o(1)$ and $\mathrm{Var}_\theta(v) = 1/(n \cdot \text{present } I) + o(1)$.

A special case of the above is when $\theta_1 = \theta_2$ – i.e., $k = 1$ – so that

$$\ell(\theta \mid s) = \theta_1^{f_{11}(s)+f_{22}(s)}(1-\theta_1)^{f_{12}(s)+f_{21}(s)} = \theta_1^{y(s)}(1-\theta_1)^{n-y(s)},$$

where of course $y = f_{11} + f_{22}$. It turns out that y is a $B(n, \theta_1)$ variable, so that $\hat{\theta}_1 = y/n$ satisfies $\mathrm{Var}_\theta(\hat{\theta}_1) = \frac{1}{n}\theta_1(1-\theta_1)$. This is the new I^{-1}.

Example 6. $X_i \sim N(0,1)$, $\Theta = (0,1)$.

a. $\text{Cov}_\theta(X_i, X_j) = \theta^{j-i}$ for all $i < j$.

$$u_1 = \frac{X_1 X_2 + X_2 X_3 + \cdots + X_{n-1} X_n}{n-1}$$

is unbiased for θ and

$$u_2 = \frac{X_1 X_3 + X_2 X_4 + \cdots + X_{n-2} X_n}{n-2}$$

is unbiased for θ^2. $u_1 + k\sqrt{u_2}$ is an estimate of θ; what are its properties?

b. $\text{Cov}_\theta(X_i, X_j) = \theta$ for all $i \neq j$.

In both cases $(X_1, \ldots, X_n)$ is from a stationary sequence. What is $I(\theta)$ in 6(a) and 6(b)? What estimate(s) t ($t = \hat{\theta}$? $t = u^*$? $t = u^{**}$?) has (have) the property that $E_\theta(t) \approx \theta$ and $\text{Var}_\theta(t) \approx I^{-1}(\theta)$ for large n?
In 6(a), find $|C|$ and C^{-1}, where

$$C = \text{Cov}_\theta(s) = \begin{pmatrix} 1 & \theta & \cdots & \theta^{n-1} \\ \theta & 1 & \ddots & \vdots \\ \vdots & \ddots & \ddots & \theta \\ \theta^{n-1} & \cdots & \theta & 1 \end{pmatrix}.$$

(C^{-1} is tridiagonal.) In 6(b), find $|D|$ and D^{-1}, where

$$D = \begin{pmatrix} 1 & \theta & \cdots & \theta \\ \theta & 1 & \ddots & \vdots \\ \vdots & \ddots & \ddots & \theta \\ \theta & \cdots & \theta & 1 \end{pmatrix}.$$

$(D = (1-\theta)I + \theta u$, where $u = \begin{pmatrix} 1 & \cdots & 1 \\ \vdots & \ddots & \vdots \\ 1 & \cdots & 1 \end{pmatrix}$, so $D^{-1} = \alpha I + \beta u$.)

Homework 5

2. (Optional) Answer the questions in Example 6.

A review of the preceding heuristics

Suppose θ is real.

i. CONSISTENCY: $\hat{\theta}$ is close to the true θ.

ii. $E_\theta(\hat{\theta}) \approx \theta$; in fact, $E_\theta(\hat{\theta}) = \theta + O(1/\sqrt{I(\theta)})$.

iii. $\operatorname{Var}_\theta(\hat\theta) = \frac{1}{I(\theta)} + o\left(\frac{1}{I(\theta)}\right)$.

If u is any estimate such that $u = \theta + O\left(\frac{1}{\sqrt{I(\theta)}}\right)$, then u^*, u^{**} etc. also have properties (ii) and (iii).

Consistency is difficult even today. Assuming that $\hat\theta$ exists and is consistent, then (ii) and (iii) remain difficult, but one can say that $\hat\theta$, u^*, u^{**}, etc. are $\approx N\left(\theta, 1/I(\theta)\right)$ where $I(\theta)$ is large.

Theorem (on consistency). *Let X_i be iid. $\ell(\theta \mid X_i)$ depends on $\theta \in \Theta = (a, b)$ with $-\infty \le a < b \le +\infty$, and $\ell(\theta \mid s) = \prod_{i=1}^n \ell(\theta \mid X_i)$.*

Condition 1. For all s, $\ell(\cdot \mid s)$ is continuous.

Let $\hat\theta_n : S \to \Theta$ be some function; $\hat\theta$ is an ML estimate $\Leftrightarrow$ $\hat\theta$ is measurable and

$$\ell\left(\hat\theta(s) \mid s\right) = \sup_{\delta \in \Theta} \ell(\delta \mid s)$$

whenever the supremum exists.

Condition 2. $\lim_{\theta \to a} \ell(\theta \mid X_1)$ and $\lim_{\theta \to b} \ell(\theta \mid X_1)$ exist a.e. with respect to the dominating measure for X_1; denote these limits by $\ell(a \mid X_1)$ and $\ell(b \mid X_1)$.

Condition 3. If $\theta \in \Theta$, then

$$\{x_1 : \ell(\theta \mid x_1) \ne \ell(a \mid x_1)\}$$

and

$$\{x_1 : \ell(\theta \mid x_1) \ne \ell(b \mid x_1)\}$$

have positive measures (with respect to the dominating measure for X_1). For any $\theta, \delta \in \overline\Theta$ with $\theta \ne \delta$,

$$\{x_1 : \ell(\theta \mid x_1) \ne \ell(\delta \mid x_1)\}$$

has positive measure.

   1 (LeCam). Condition 1 implies that an ML estimate exists.

   2 (Wald). Conditions 1–3 imply that, for all $\theta \in \Theta$, with probability 1,

      1. $\hat\theta_n$ actually maximizes the likelihood for all sufficiently large n.

      2. $\lim_{n \to \infty} \hat\theta_n = \theta$.

Note. The proof of (2) depends on the fact that $[a, b]$ is compact. There are difficulties in extending the proof to, say, $\Theta \subseteq \mathbb{R}^p$, because it is difficult to find a suitable compactification of Θ.

Chapter 8

Lecture 28

Example 7. X_i are iid uniformly over $(0, \theta)$ for $\theta \in \Theta = (0, \infty)$.

Homework 6

1. Show that

   a. With respect to Lebesgue measure on $\mathbb{R}^n$,

   $$\ell(\theta \mid s_n) = \begin{cases} 1/\theta^n & \text{if } \theta \geq X_i \ \forall i \\ 0 & \text{otherwise} \end{cases}$$

   and $\hat{\theta} = \max\{X_1, \ldots, X_n\}$.

   b. Condition 2 in the Theorem above is satisfied, and hence $\hat{\theta}_n \xrightarrow{\text{a.s.}} \theta$ for all θ (which we check directly also); but the likelihood function is not continuous, and hence the information function is not defined.

   c. $E_\theta(\hat{\theta}_n) = \frac{n}{n+1}\theta$, and $\theta_n^* := \frac{n+1}{n}\hat{\theta}$ is unbiased.

   d. $n(\theta - \hat{\theta}_n)$ has the asymptotic distribution with density $\frac{1}{\theta}e^{-\frac{z}{\theta}}$ on $(0, \infty)$, and so $\hat{\theta}_n$ has a non-normal limiting distribution and $\hat{\theta}_n - \theta = O(1/n)$.

   (In regular cases, $\hat{\theta}$ has a normal limiting distribution and $\hat{\theta}_n - \theta = O(1/\sqrt{n})$.)

Asymptotic distribution of $\hat{\theta}$ (θ real) in regular cases

$X = \{x\}$ (arbitrary), $\mathcal{C}$ is a σ-field on X, P_θ is a probability on $\mathcal{C}$ and $\theta \in \Theta$ for Θ an open interval in $\mathbb{R}^1$. $dP_\theta(x) = \ell(\theta \mid x)d\nu(x)$, with ν a fixed measure. Let $s_n = (X_1, \ldots, X_n) \in S^{(n)} = X \times \cdots \times X$, $\mathcal{A}^{(n)} = \mathcal{C} \times \cdots \times \mathcal{C}$ and $P_\theta^{(n)} = P_\theta \times \cdots \times P_\theta$ on $\mathcal{A}^{(n)}$. We assume that $\ell(\theta \mid x) > 0$, $L(\theta \mid x) = \log_e \ell(\theta \mid x)$ has at least two continuous derivatives, $E_\theta(L'(\theta \mid x)) = 0$ and

$$I_1(\theta) = E_\theta(L'(\theta \mid x))^2 = -E_\theta(L''(\theta \mid x)) > 0.$$

We have $L(\theta \mid s_n) = \sum_{i=1}^{n} L(\theta \mid X_i)$, $L'(\theta \mid s_n) = \sum_{i=1}^{n} L'(\theta \mid X_i)$ and $L''(\theta \mid s_n) = \sum_{i=1}^{n} L''(\theta \mid X_i)$. For any given θ, we know that a good estimate of θ based on s_n will be approximately $a(\theta) + b(\theta)L'(\theta \mid s_n)$, and $L'(\theta \mid s_n) \approx N(0, *)$, so a good estimate of θ based on s_n will be approximately normally distributed when n is large. We have $\frac{L''(\theta \mid s_n)}{n} \to -I_1(\theta)$. Assume that:

Condition (*). Given any $\theta \in \Theta$, we may find an $\varepsilon = \varepsilon(\theta) > 0$ such that

$$\max_{|\delta - \theta| \le \varepsilon} |L''(\delta \mid x)|$$

has a finite expectation under P_θ.

Assume also that $\hat{\theta}_n$ exists and is consistent. Then

$$0 = L'(\hat{\theta}_n \mid s_n) = L'(\theta \mid s_n) + (\hat{\theta}_n - \theta)L''(\theta_n^* \mid s_n),$$

where θ_n^* is between θ and $\hat{\theta}_n$. Since $\theta_n^* \to \theta$ in P_θ, we have

$$\left| \frac{L''(\theta_n^* \mid s_n)}{n} + I_1(\theta) \right| \to 0 \quad \text{in } P_\theta. \qquad (**)$$

So

$$\sqrt{n}(\hat{\theta}_n - \theta) = \frac{L'(\theta \mid s_n)}{\sqrt{n}} \cdot \frac{1}{I_1(\theta) + \xi_n},$$

where $\xi_n \to 0$ in P_θ. Since

$$\frac{L'(\theta \mid s_n)}{\sqrt{n}} \to N\big(0, I_1(\theta)\big) \quad \text{in distribution under } P_\theta,$$

we have:

1 (Fisher). $\sqrt{n}(\hat{\theta}_n - \theta) \to N\big(0, I_1(\theta)\big)$.

Note. This does *not* assert that $E_\theta(\hat{\theta}_n) = \theta + o(1)$ or that $\text{Var}_\theta(\hat{\theta}_n) = \frac{1}{nI_1(\theta)} + o\big(\frac{1}{n}\big)$.

Proof of (**). Fix θ. Under (*), we have

$$h(r) := E_\theta\Big[\max_{|\delta - \theta| \le r} |L''(\delta \mid x) - L''(\theta \mid x)|\Big] < +\infty$$

for sufficiently small $r > 0$. h is continuous in r and decreases to 0 as $r \to 0$.

For any $\eta > 0$, choose r such that $h(r) < \eta$. We have

$$\frac{1}{n}L''(\theta_n^* \mid s_n) = \frac{1}{n}L''(\theta \mid s_n) + \Delta_n,$$

where

$$|\Delta_n| = \frac{1}{n}\left|\sum_{i=1}^{n}[L''(\theta_n^* \mid X_i) - L''(\theta \mid X_i)]\right| \le \frac{1}{n}\sum_{i=1}^{n}|L''(\theta_n^* \mid X_i) - L''(\theta \mid X_i)|.$$

68

Suppose that $|\hat{\theta}_n - \theta| < r$; then $|\theta_n^* - \theta| < r$ and hence $|\Delta_n| \leq \frac{1}{n}\sum_{i=1}^n M(X_i)$, where $M(X) = \max_{\delta - \theta| \leq r}|L''(\delta \mid X) - L''(\theta \mid X)|$.

Since $E[M(X_i)] < \eta$, we have

$$\frac{1}{n}\sum_{i=1}^n M(X_i) \xrightarrow{\text{a.s.}} E_\theta[M(X)] < \eta.$$

Since η is arbitrary and $\hat{\theta}_n \to \theta$ in P_θ, we have that $|\Delta_n| \to 0$ in P_θ. $\square$

Note. It was asserted by Fisher (and believed for a long time) that, if $t_n = t_n(s_n)$ is any estimate of θ such that

$$\sqrt{n}(t_n - \theta) \to N(0, v(\theta)) \quad \text{in distribution as } n \to \infty,$$

then $v(\theta) \geq 1/I_1(\theta)$. This is, however, not quite correct, as shown by the following counterexample (due to J. L. Hodges, 1951): Let X_i be iid $N(\theta, 1)$ and $\Theta = \mathbb{R}^1$. Let $\hat{\theta}_n = \overline{X}_n$. $\sqrt{n}(\hat{\theta}_n - \theta)$ is $N(0,1)$ and $I_1(\theta) = 1$. Let

$$t_n = \begin{cases} \overline{X}_n & \text{if } |\overline{X}_n| > n^{-1/4} \\ c\overline{X}_n & \text{if } |\overline{X}_n| \leq n^{-1/4}; \end{cases}$$

then $\sqrt{n}(t_n - \theta) \to N(0, v(\theta))$ for all θ, where

$$v(\theta) = \begin{cases} 1 & \text{if } \theta \neq 0 \\ c^2 & \text{if } \theta = 0, \end{cases}$$

and so $v(\theta) \geq 1$ breaks down at $\theta = 0$ (if we choose $-1 < c < 1$).

Lecture 29

Definition. We say that $\{z_n\}$ is $AN(\mu_n, \sigma_n^2)$ if

$$P\left(\frac{z_n - \mu_n}{\sigma_n} \leq z\right) \to \Phi(z) \quad \text{for all } z.$$

Consider the condition

Condition (***). $\{t_n - \theta\}$ is $AN(0, v(\theta)/n)$ under θ (for each θ).

In Hodges's counterexample in the context of Example 1(a),

$$\sqrt{n}(t_n - \theta) = \varphi(\theta)\sqrt{n}(\overline{X}_n - \theta) + \xi_n(s, \theta),$$

where $\xi_n \to 0$ in P_θ-probability and

$$\varphi(\theta) = \begin{cases} 1 & \text{if } \theta \neq 0 \\ c & \text{if } \theta = 0, \end{cases}$$

so that t_n is $AN(0, v(\theta)/n)$ for $v(\theta) = \varphi^2(\theta)$. This provides an example of the following theorem:

2 (Le Cam/Bahadur). The set

$$\left\{\theta : v(\theta) < \frac{1}{I_1(\theta)}\right\}$$

is always of Lebesgue measure zero for any t_n satisfying (***).

Corollary. *If $\{t_n\}$ is regular in the sense that v is continuous in Θ and I_1 is also continuous, then $v(\theta) \geq 1/I_1(\theta)$ for all $\theta \in \Theta$.*

Note. This should not be confused with the C-R bound, since (***) does *not* imply that t_n is unbiased, nor that $v(\theta) \cong n \operatorname{Var}_\theta(t_n)$.

In the general case, (***) does imply that t_n is asymptotically median unbiased, i.e., that $P_\theta(t_n \leq \theta) \to \frac{1}{2}$ as $n \to \infty$ for each θ. Suppose this holds uniformly; then also it must be true that $v(\theta) \geq 1/I_1(\theta)$ for all θ. This follows from:

3. If θ is a point in Θ, $a > 0$ and $\delta_n(a) = \theta + \frac{a}{\sqrt{n}}$, and

$$\varliminf_{n\to\infty} P_{\delta_n(a)}\bigl(t_n > \delta_n(a)\bigr) \geq \frac{1}{2},$$

then $v(\theta) \geq 1/I_1(\theta)$ (for the given θ).

Corollary. *Suppose that t_n is super-efficient $(v < 1/I_1)$ at a point θ. Then, given any $a > 0$, we may find $\varepsilon_1 = \varepsilon_1(a) > 0$ and $\varepsilon_2 = \varepsilon_2(a) > 0$ such that*

$$P_{\theta + \frac{a}{\sqrt{n}}}\left(t_n > \theta + \frac{a}{\sqrt{n}}\right) < \frac{1}{2} - \varepsilon_1 \quad and \quad P_{\theta - \frac{a}{\sqrt{n}}}\left(t_n < \theta - \frac{a}{\sqrt{n}}\right) < \frac{1}{2} - \varepsilon_2$$

for all sufficiently large n.

Definition. Let F_n be a sequence of distributions on $\mathbb{R}^k$ and F_0 be a given distribution on $\mathbb{R}^k$. We say that $F_n \xrightarrow{\mathcal{L}} F_0$ iff

$$\int_{\mathbb{R}^k} b(x)dF_n(x) \to \int_{\mathbb{R}^k} b(x)dF_\theta(x)$$

for all bounded continuous functions $b : \mathbb{R}^k \to \mathbb{R}^1$.

4 (Hájek). Let $F_{n,\theta} = \mathcal{L}\bigl(\sqrt{n}(\tau_n - \theta)\bigr)$ and suppose that $F_{n,\theta + \frac{a}{\sqrt{n}}} \xrightarrow{\mathcal{L}} G$ for all $|a| \leq 1$. Then G is the distribution function of $X + Y$, where X is $N\bigl(0, 1/I_1(\theta)\bigr)$ and X and Y are independent. (This is true for all θ. G can depend on θ.)

Corollary. *The variance of G (if it exists) is at least $1/I_1(\theta)$.*

Conclusion. At least in the iid case, Fisher's assertion is essentially correct.

Proof of (3) (outline). Choose $\theta \in \Theta$ and $a > 0$, and let $\delta_n = \theta + \frac{a}{\sqrt{n}}$. For fixed n, consider testing θ against δ_n. $\frac{\ell(\delta_n|s_n)}{\ell(\theta|s_n)}$ is the optimal (LR) test statistic, whose logarithm is

$$L_n(\delta_n) - L(\theta) = \frac{a}{\sqrt{n}}L'(\theta) + \frac{a^2}{2n}L''(\theta_n^*) = \frac{a}{\sqrt{n}}L'(\theta) - \frac{1}{2}a^2 I_1(\theta) + \cdots,$$

where the omitted terms are negligible. Let

$$K_n(s_n) = \frac{1}{\sqrt{a^2 I_1(\theta)}}\left(L(\delta_n \mid s_n) - L(\theta \mid s_n) + \frac{1}{2}a^2 I_1(\theta) \right).$$

K_n is equivalent to the LR statistic and $K_n \xrightarrow{\mathcal{L}} N(0,1)$ under P_θ. Consider the distribution of K_n under δ_n,

$$P_{\delta_n}(K_n < z) = \int_{K_n < z} dP_{\delta_n}^{(n)} = \int_{K_n(s_n) < z} e^{L(\delta_n|s_n) - L(\theta|s_n)} dP_\theta^{(n)}(s_n)$$

$$= \int_{K_n(s_n) < z} e^{-\frac{1}{2}a^2 I_1(\theta) + \sqrt{a^2 I_1(\theta)} K_n(s_n)} dP_\theta^{(n)}(s_n) = \int_{y < z} e^{-\frac{1}{2}a^2 I_1(\theta) + \sqrt{a^2 I_1(\theta)} y} dF_n(y)$$

$$\to \int_{y < z} e^{-\frac{1}{2}a^2 I_1(\theta) + \sqrt{a^2 I_1(\theta)} y} d\Phi(y) = \frac{1}{\sqrt{2\pi}} \int_{-\infty}^{z} e^{-\frac{1}{2}a^2 I_1(\theta) + \sqrt{a^2 I_1(\theta)} y - \frac{1}{2}y^2} dy$$

$$= P\big(N(0,1) < z - \sqrt{a_2 I_1(\theta)}\big),$$

where $F_n(y) = P_\theta(K_n < y)$. Note that $F_n(y) \to \Phi(y)$.

Given a sequence $\{t_n\}$ such that $\overline{\lim}_{n\to\infty} P_{\delta_n}(t_n \geq \delta_n) \geq 1/2$, choose $z > \sqrt{a^2 I_1(\theta)}$. Then, by the above result, $P_{\delta_n}(K_n \geq z) < 1/2$ for all sufficiently large n. Regard $\{t_n \geq \delta_n\}$ and $\{K_n \geq z_n\}$ as critical regions for the test; then, by the Neyman-Pearson lemma, we have that, for some subsequence $\{n_k\}$, $P_\theta(K_{n_k} > z) \leq P_\theta(t_{n_k} \geq \delta_{n_k})$ for all sufficiently large k; but

$$P_\theta(t_n \geq \delta_n) = P_\theta\big(\sqrt{n}(t_n - \theta) \geq a\big) \quad \text{and} \quad P_\theta(K_n \geq z) \to 1 - \Phi(z),$$

so

$$z > \sqrt{a^2 I_1(\theta)} \Rightarrow P_\theta(K_{n_k} > z) < P_\theta(t_{n_k} \geq \theta + a/\sqrt{n_k}).$$

Letting $k \to \infty$, we find that

$$P\big(N(0,1) \geq z\big) \leq P\big(N(0,1) \geq a/\sqrt{v(\theta)}\big)$$

and hence $z > a/\sqrt{v(\theta)}$. Since z was arbitrary, we must have $\sqrt{a^2 I_1(\theta)} \geq a/\sqrt{v(\theta)}$ and hence $v(\theta) \geq 1/I_1(\theta)$. $\qquad\square$

Lecture 30

Proof of (2). Assume only (***), i.e., that $\sqrt{n}(t_n - \theta) \xrightarrow{\mathcal{L}_\theta} N\big(0, v(\theta)\big)$ for $\theta \in \Theta$, and let J be a bounded subinterval of Θ, say (a, b). Let

$$\Psi_n(\theta) = P_\theta(t_n > \theta) \quad \text{and} \quad \varphi_n(\theta) = \left| \Psi_n(\theta) - \frac{1}{2} \right|.$$

Then $0 \le \varphi_n(\theta) \le \frac{1}{2}$ and, from (***), $\Psi_n(\theta) \to \frac{1}{2}$ and $\varphi_n(\theta) \to 0$ for each θ. Hence $\theta \mapsto I_J(\theta)\varphi_n(\theta)$, where I_J is an indicator function, is bounded on Θ and tends to 0, so $\int_\Theta I_J(\theta)\varphi_n(\theta)d\theta \to 0$, or

$$\int_{\mathbb{R}^1} I_J\left(\delta + \frac{1}{\sqrt{n}}\right) \varphi_n\left(\delta + \frac{1}{\sqrt{n}}\right) d\delta \to 0;$$

but $I_J(\delta + \frac{1}{\sqrt{n}}) \to I_J(\delta)$ except for δ an endpoint of J, so

$$\int_{\mathbb{R}^1} I_J(\delta)\varphi_n\left(\delta + \frac{1}{\sqrt{n}}\right) d\delta \to 0.$$

Noticing that $I_J(\delta)\varphi_n\big(\delta + \frac{1}{\sqrt{n}}\big) \ge 0$, we have $I_J(\delta)\varphi_n\big(\delta + \frac{1}{\sqrt{n}}\big) \to 0$ in Lebesgue measure, so that there is some sequence $\{n_k\}$ such that $I_J(\delta)\varphi_{n_k}\big(\delta + \frac{1}{\sqrt{n_k}}\big) \to 0$ a.e.(Lebesgue); thus $\varphi_{n_k}\big(\delta + \frac{1}{\sqrt{n_k}}\big) \to 0$ a.e.(Lebesgue) on J – i.e., $P_{\theta + \frac{1}{\sqrt{n_k}}}\big(t_{n_k} > \theta + \frac{1}{\sqrt{n_k}}\big) - \frac{1}{2} \to 0$ a.e. on J. Returning to the original sequence, we have that $\overline{\lim}_{n \to \infty} P_{\theta + \frac{1}{\sqrt{n}}}\big(t_n > \theta + \frac{1}{\sqrt{n}}\big) \ge 1/2$ a.e. on J and so, from (3), $v(\theta) \ge 1/I_1(\theta)$ a.e. on J. Since J was *any* bounded subinterval of Θ, this means that $v(\theta) \ge 1/I_1(\theta)$ a.e. on Θ. $\qquad\square$

General regular case

For each n, let $(S_n, \mathcal{A}_n, P_\theta^{(n)})$ be an experiment with common parameter

$$\theta = (\theta_1, \ldots, \theta_p) \in \Theta,$$

where Θ is open in $\mathbb{R}^p$, such that S_n consists of points s_n. No relation between n and $n+1$ is assumed.

In Examples 1–5, we have $S_n = \underbrace{X \times \cdots \times X}_{n \text{ times}}$ and $P_\theta^{(n)} = P_\theta \times \cdots P_\theta$. In Examples 6 and 7, $P_\theta^{(n)}$ is the distribution of $s_n = (X_1, \ldots, X_n)$, where the X_i are *not* iid.

Example 8. For $n = 2, 3, \ldots$, let n_1 and n_2 be positive integers such that $n = n_1 + n_2$. Let $s_n = (X_1, \ldots, X_{n_1}; Y_1, \ldots, Y_{n_2})$, where $X_1, \ldots, X_{n_1}, Y_1, \ldots, Y_{n_2}$ are independent, $X_1, \ldots, X_{n_1}$ are $N(\mu_1, \sigma^2)$ distributed and $Y_1, \ldots, Y_{n_2}$ are $N(\mu_2, \sigma^2)$ distributed. Here $\theta = (\mu_1, \mu_2, \sigma^2)$ is entirely unknown. This is a three-parameter exponential family, and the complete sufficient statistic is

$$\left(\sum_{i=1}^{n_1} X_i, \sum_{i=1}^{n_2} Y_i, \sum_{i=1}^{n_1} X_i^2 + \sum_{i=1}^{n_2} Y_i^2 \right).$$

If $n_1/n_2 \to \rho$ as $n \to \infty$ for some $0 < \rho < \infty$, all regularity conditions to follow are satisfied.

The local asymptotic normality condition

Choose $\theta \in \Theta$ and assume that $dP_\delta^{(n)}(s_n) = \Omega_{\delta,\theta}(s_n)dP_\theta^{(n)}(s_n)$ holds for all δ in a neighborhood of θ.

Condition LAN (at $\theta \in \Theta$). For each $a \in \mathbb{R}^p$,

$$\log_e\left(\Omega_{\theta+\frac{a}{\sqrt{n}},\theta}(s_n)\right) = az_n'(\theta) - \frac{1}{2}a'I_1(\theta)a + \Delta_n(\theta, s_n),$$

where I_1 is a fixed $p \times p$ positive definite matrix, $z_n(\theta) \in \mathbb{R}^p$ and $z_n(\theta) \xrightarrow{\mathcal{L}_\theta} N\left(0, I_1(\theta)\right)$ and $\Delta_n(\theta, s_n) \to 0$ in $P_\theta^{(n)}$-probability.

Note.

i. If $s_n = (X_1, \ldots, X_n)$, where the X_is are iid, and I_1 is the information matrix for X_1, then LAN is satisfied for this I_1; but the LAN condition holds in some "irregular" cases also – see Example 1(b).

ii. The right-hand side in LAN with Δ_n omitted is exactly the log-likelihood in the multivariate normal translation-parameter case. See Example 4.

Let $g : \Theta \to \mathbb{R}^1$ be continuously differentiable and write $h(\theta) = \operatorname{grad} g(\theta)$.

2^p (Le Cam). If $t_n = t_n(s_n)$ is an estimate of g such that

$$\sqrt{n}\left(t_n - g(\theta)\right) \xrightarrow{\mathcal{L}_\theta} N\left(0, v(\theta)\right) \; \forall \theta \in \Theta,$$

then $\{\theta : v(\theta) < b_1(\theta)\}$ is of (p-dimensional) Lebesgue measure 0 if we let $b_1(\theta) = h(\theta)I_1^{-1}(\theta)h'(\theta)$.

4^p (Hájek). Suppose that $u_n : S_n \to \Theta$ is s.t.

$$\sqrt{n}\left(u_n - (\theta + a/\sqrt{n})\right) \xrightarrow{\mathcal{L}_{\theta+a/\sqrt{n}}} u_\theta$$

(u_θ independent of a), then u_θ may be represented as $v_\theta + w_\theta$, where v_θ and w_θ are independent and $v_\theta \sim N\left(0, I_1^{-1}(\theta)\right)$.

Note. No uniformity in a is needed in Hájek's theorem.

From the above we see that, for large n, the $N\left(0, I_1^{-1}(\theta)/n\right)$ distribution is nearly the best possible for estimates of θ. n is the "sample size", or cost of observing s_n.

Sufficient conditions for LAN

Suppose that $L(\theta \mid s_n)$ exists for each n, i.e., that $dP_\theta^{(n)}(s_n) = e^{L(\theta \mid s_n)} d\nu^{(n)}(s_n)$ for all n, and that, for each n, $L(\cdot \mid s_n)$ has at least two continuous derivatives. We write $\ell = e^L$. Let $L^{(1)}(\theta \mid s_n) = \operatorname{grad} L(\theta \mid s_n)$.

Condition 1. $\frac{1}{\sqrt{n}} L^{(1)}(\theta \mid s_n) \xrightarrow{\mathcal{L}_\theta} N(0, I_1(\theta))$ for some positive definite I_1.

Condition 2. $\frac{1}{n}\{L_{ij}(\theta \mid s_n)\} \to -I_1(\theta)$ in $P_\theta^{(n)}$-probability.

Condition 3. With

$$M(\theta, \gamma, s_n) := \frac{1}{n} \max_{\substack{\|\delta - \theta\| \leq \gamma \\ i,j=1,\ldots,p}} \{|L_{ij}(\delta \mid s_n) - L_{ij}(\theta \mid s_n)|\},$$

$\lim_{r \downarrow 0} \overline{\lim}_{n \to \infty} P_\theta^{(n)}(M(\theta, \gamma, s_n) > \varepsilon) = 0$ for every $\varepsilon > 0$.

Conditions 1–3 imply LAN with $\Delta_n \to 0$, and also the following:

1^p (Fisher). Under Conditions 1–3, if $\hat{\theta}_n = \hat{\theta}_n(s_n)$, the MLE of θ, exists and is consistent, then

$$\sqrt{n}(\hat{\theta}_n - \theta) \xrightarrow{\mathcal{L}_\theta} N(0, I_1^{-1}(\theta)) \quad \forall \theta \in \Theta.$$

Definition. Let $u_n = u_n(s_n)$ be an estimate of θ. u_n is CONSISTENT if $u_n \xrightarrow{P_\theta} \theta$ for all θ, or, equivalently, $(u_n - \theta)(u_n - \theta)' \xrightarrow{P_\theta} 0$. u_n is $\sqrt{n}$-CONSISTENT if $n(u_n - \theta)(u_n - \theta)'$ is bounded in P_θ for all θ. (We say that Y_n is BOUNDED in P if, given any $\varepsilon > 0$, we may find k such that $P(|Y_n| > k) \leq \varepsilon$ for all n sufficiently large.)

1^p (continued). If u_n is a $\sqrt{n}$-consistent estimate of θ and

$$u_n^* = u_n + \left\{ (L_{ij}(\theta \mid s_n))^{-1} L^{(1)}(\theta \mid s_n) \big|_{\theta = u_n} \right\}$$

and

$$u_n^{**} = u_n + \left\{ I_n(\hat{\theta}_n)^{-1} L^{(1)}(\theta \mid s_n) \big|_{\theta = u_n} \right\},$$

then u_n^* and u_n^{**} are both $AN(\theta, I_1^{-1}(\theta)/n)$. Consequently, $t_n^* = g(u_n^*)$ and $t_n^{**} = g(u_n^{**})$ are both $AN(g(\theta), b_1(\theta)/n)$, where $b_1(\theta) = h(\theta) I_1^{-1}(\theta) h'(\theta)$.

References

Chandrasekhar, S. (1995). *Newton's Principia for the Common Reader*. Oxford: Oxford University Press.

Cramér, H. (1946). *Mathematical Methods of Statistics*. Princeton: Princeton University Press.

Ibragimov, I. A., and R. Z. Has'minskii (1981). *Statistical Estimation: Asymptotic Theory*. New York: Springer-Verlag.

Joshi, V. M. (1976). On the Attainment of the Cramér-Rao Lower Bound. *Annals of Statistics*, **4**, 998-1002.

Le Cam, L. (1953). On Some Asymptotic Properties of Maximum Likelihood Estimates and Related Bayes's Estimates. *University of California Publications in Statistics*, **1**, 277-330.

Lehmann, E. L. (1991). *Theory of Point Estimation*, 2nd ed. (1st ed. 1983). Pacific Grove: Wadsworth.

Pitman, E. J. G. (1979). *Some Basic Theory of Statistical Inference*, [2.1] in Chapter Two. Pacific Grove: Wadsworth.

Rao, C. R. (1973). *Linear Statistical Inference*. New York: John Wiley.

Stuart, A., J. K. Ord and S. Arnold (1994). *Kendall's Advanced Theory of Statistics* Chapter 10. London: Arnold.

Wijsman, R. A. (1973). On the Attainment of the Cramér-Rao Lower Bound. *Annals of Statistics*, **1**, 538-542.

Wong, W. H. (1992). On Asymptotic Efficiency in Estimation Theory. *Statistica Sinica*, **2**, 47-68.

Zheng, Z. and K. T. Fang (1994). On Fisher's Bound for Stable Estimators with Extension to the Case of Hilbert Parameter Space. *Statistica Sinica*, **4**, 679-692.

Raghu Raj Bahadur 1924-1997

By Stephen M. Stigler

Raghu Raj Bahadur was born in Delhi, India on April 30, 1924. He was extremely modest in demeanor and uncomfortable when being honored, but on several occasions the Chicago Department of Statistics managed to attract him to birthday celebrations by taking advantage of the coincidence of his and Gauss's birthdates -- he would come to honor Gauss, n to receive honor for himself. At St. Stephen's College of Delhi University he excelled, graduating in 1943 with first class honors in mathematics. In 1944 he won a scholarship and generously returned the money to the College to aid poor students. But it is clear that he had n yet found his calling. In that year, 1944, his was judged the best serious essay by a student in t College. The essay gave no hint of the career that was to follow: it was a somber essay, on the isolation of individuals, and it gave a dark and pessimistic view of the search for meaning in lif – a vision that was foreign to the Raj we knew in later years. He continued on at Delhi, receivi a Masters degree in mathematics in 1945. After a year at the Indian Institute of Science in Bangalore he was awarded a scholarship by the government of India for graduate studies, and i October 1947, after spending one year at the Indian Statistical Institute in Calcutta, Raj took an unusual and fateful step. While India was in the upheaval that followed partition and preceded independence, Raj traveled to Chapel Hill, North Carolina, to study mathematical statistics.

In barely over two years he completed his Ph.D. His dissertation focused on decision theoretic problems for k populations, a problem suggested by Harold Hotelling (although Herbe Robbins served as his major professor). In a December 1949 letter of reference, Harold Hotelling wrote: "His thesis, which is now practically complete, includes for one thing a discussion of the following paradox: Two samples are known to be from Cauchy populations

ose central values are known, but it is not known which is which. Probability of erroneous
ignment of the samples to the two populations may be larger in some cases when the greater
nple mean is ascribed to the greater population mean than when the opposite is done." His
st paper, including this example, was published in the *Annals of Mathematical Statistics* (1).

At the winter statistical meetings in December 1949, W. Allen Wallis contacted him to
and him out – was he interested in joining the new group of statisticians being formed at the
iversity of Chicago? He was interested, and Allen arranged for Raj to start in the Spring
arter of 1950. Raj's move to Chicago was to prove a pivotal event in his life. He left Chicago
ice (in 1952 and in 1956), and he returned twice (in 1954 and in 1961). He never forgot his
ots in India, and the pull of family and the intellectual community in Delhi caused him to
urn there time and again throughout his life, but Chicago had a special, irresistible allure for
n. In the decade following 1948, Allen Wallis assembled an extraordinarily exciting and
fluential intellectual community. Starting with Jimmie Savage, William Kruskal, Leo
oodman, Charles Stein, and Raj Bahadur, he soon added David Wallace, Paul Meier, and
trick Billingsley.

Raj thrived at Chicago, although sometimes the pric.e was high. One of his great
hievements was his 1954 paper on "Sufficiency and Statistical Decision Functions," (5) a
onumental paper (it ran to 40 pages in the *Annals*) that is a masterpiece of both mathematics
d statistics. The story of its publication tells much about the atmosphere in Chicago in those
ys. It was originally submitted in May of 1952, and, with Raj away in India, it was assigned to
nmie Savage as a referee. Savage was favorable and impressed, and fairly quick in his report
e took two months on what must have been a 100 page manuscript of dense mathematics), so
ny was there a two year delay in publication? It was not because of a backlog; the *Annals* was

publishing with a three month delay in those days. Rather it was the character of the report and the care of Raj's response. For while Savage was favorable, his reports (eventually there were three) ran to 20 single-spaced pages, asking probing questions as well as listing over 60 points linguistic and mathematical style. Somehow Raj survived this barrage, rewriting the paper completely, benefiting from the comments but keeping the work his own, and preserving, over another referee's objections, an expository style that explained the deep results both as mathematics and again as statistics.

From 1956-61 Raj was again in India, this time as a Research Statistician at the Indian Statistical Institute, Calcutta, but in 1961 he returned to the University of Chicago to stay, except for two leaves back to India. He retired in 1991 but continued to take vigorous part in the intellectual life of the Department as long as his increasingly frail health permitted. He died on June 7, 1997.

Bahadur's research in the 1950s and 1960s played a fundamental role in the development of mathematical statistics over that period. These works included a series of papers on sufficiency (5)-(8), investigations on the conditions under which maximum likelihood estimate will be consistent (including Bahadur's Example of Inconsistency) (12), new methods for the comparison of statistical tests (including the measure based upon the theory of large deviations now known as Bahadur Efficiency) (16,17), and an approach to the asymptotic theory of quantiles (now recognized as the Bahadur Representation of Sample Quantiles) (25). C. R. Rao has written "Bahadur's theorem [his 1957 converse to the Rao-Blackwell theorem (11)] is one the most beautiful theorems of mathematical statistics" [in *Glimpses of India's Statistical Heritage*, Ed. J. K. Ghosh, S. K. Mitra, K. R. Parthasarathy, Wiley Eastern, 1992, p. 162]. Other work included his approach to classification of responses from dichotomous estionnaires (including the Bahadur-Lazarsfeld Expansion) (20, 21), and the asymptotic timality of the likelihood ratio test in a large deviation sense (24). Bahadur summarized his earch in the theory of large deviations in an elegant short monograph, *Some Limit Theorems ii tistics* (30).

Virtually everything Raj did was characterized by a singular depth and elegance. He tool rticular pleasure in showing how simplicity and greater generality could be allies rather than tagonists, as in his demonstration that LeCam's theorem on Fisher's bound for asymptotic riances could be derived from a clever appeal to the Neyman-Pearson Lemma (23). Raj's ork was remarkable for its elegance and deceptive simplicity. He forever sought the "right" y of approaching a subject – a combination of concept and technique that not only yielded the ult but also showed precisely how far analysis could go. Isaac Newton labored hard to draw e right diagram, to outline in simple steps a demonstration that made the most deep and subtle inciples of celestial mechanics seem clear and unavoidably natural. Raj had a similar touch in athematical statistics. His own referee's reports were minor works of art; his papers often asterpieces.

In the early 1950s he married Thelma Clark, and together they raised two fine children, khar and Sheila Ann, of whom they were immensely proud. From his first arrival in 1950 for e rest of his life, Raj felt Chicago was a precious place. He evidently found here, in the ellectual life of the Department and the close companionship of his family, the meaning he d been seeking when he wrote his somber essay in 1944.

Raj Bahadur was President of the IMS in 1974-75, and he was the IMS's 1974 Wald cturer. He was honored by the Indian Society for Probability and Statistics in November

1987. In 1993 a Festschrift was published in his honor, *Statistics and Probability*, edited by J. K. Ghosh, S. K. Mitra, K. R. Parthasarathy, and B. L. S. Prakasa Rao (Wiley Eastern).

Raghu Raj Bahadur

Born April 30, 1924, Delhi, India; died June 7, 1977, Chicago, Illinois.

Education

B.A. (Honours)	Mathematics (with Physics)	Delhi University (St. Stephen's College)	194
M.A.	Mathematics	Delhi University (St. Stephen's College)	194
Ph.D.	Mathematical Statistics	North Carolina University (Chapel Hill)	195

Professional Career

Research Associate in Applied Statistics, Indian Statistical Institute, Calcutta, 1946-47.
Research Associate in Mathematical Statistics, University of North Carolina, 1949-50.
Instructor in Statistics, University of Chicago, 1950-51.
Professor of Statistics, Indian Council of Agricultural Research, New Delhi, 1952-53.
Visiting Assistant Professor of Mathematical Statistics, Columbia University, 1953-54.
Assistant Professor of Statistics, University of Chicago, 1954-56.
Research Statistician, Indian Statistical Institute, Calcutta, 1956-61.
Associate Professor of Statistics, University of Chicago, 1961-65.
Professor of Statistics, University of Chicago, 1965-91.
Distinguished Visiting Professor, Indian Statistical Institute, 1972-97.
Professor Emeritus, University of Chicago, 1992-97.

Professional Memberships

Fellow, Institute of Mathematical Statistics.
Member, International Statistical Institute.
Fellow, Indian National Sciences Academy.
Fellow, Indian Academy of Sciences.

Professional Activities and Honors

John Simon Guggenheim Fellow, 1968-69.
Ten lectures on limit theorems in statistics at the SIAM regional conference at Tallahassee, Florida, in 1969.
Associate Editor, *Annals of Mathematical Statistics,* 1964-1973.
Member, Council of the Indian Statistical Institute, 1972-74.
Member, Editorial Board of *Sankhya.*
Wald Lecturer, 1974 Annual Meeting of the Institute of Mathematical Statistics.
President, Institute of Mathematical Statistics, 1974-75.

invited to Department of Mathematics, University of Maryland to deliver six lectures (September 1975) as part of their "Year in Probability and Statistics" program.
Chairman of Editorial Board of the IMS-University of Chicago Monograph Series, from April 1977.
Fellow, American Academy of Arts and Sciences, from 1986.
Outstanding Statistician of the Year, Chicago Chapter of the American Statistical Association, 1992.

Publications

1) "On a problem in the theory of k populations," *Ann. Math. Statist. 21* (1950), 362-375.
2) "The problem of the greater mean" (with H. Robbins), *Ann. Math. Statist. 21* (1950), 469-487.
3) "A property of the *t* statistic," *Sankhya 12* (1952), 78-88.
4) "Impartial decision rules and sufficient statistics" (with Leo A. Goodman), *Ann. Math. Statist. 23* (1952), 553-562.
5) "Sufficiency and statistical decision functions," *Ann. Math. Statist. 25* (1954), 423-462.
6) "Two comments on sufficiency and statistical decision functions" (with E. L. Lehmann), *Ann. Math. Statist. 26* (1955), 139-142.
7) "A characterization of sufficiency," *Ann. Math. Statist. 26* (1955), 286-293.
8) "Statistics and subfields," *Ann. Math. Statist. 26* (1955), 490-497.
9) "Measurable subspaces and subalgebras," *Proc. Amer. Math. Soc. 6* (1955), 565-570.
10) "The nonexistence of certain statistical procedures in non-parametric problems" (with L. J. Savage), *Ann. Math. Statist. 27* (1956), 1115-1122.
11) "On unbiased estimates of uniformly minimum *variance,"* *Sankhya 18* (1957), 211-224.
12) "Examples of inconsistency of maximum likelihood estimates," *Sankhya 20* (1958), 207-210.
13) "A note on the fundamental identity of sequential analysis," *Ann. Math. Statist. 29* (1958), 534-543.
14) "Some approximations to the binomial distribution function," *Ann. Math. Statist. 31* (1960), 43-54.
15) "Simultaneous comparison of the optimum and sign tests of a normal mean," *Contributions to Probability and Statistics: Essays in Honor of Harold Hotelling,* Stanford University Press, (1960), 79-88.
16) "Stochastic comparison of tests," *Ann. Math. Statist. 31* (1960), 276-295.
17) "On the asymptotic efficiency of tests and estimates," *Sankhya 22* (1960), 229-252.
18) "On deviations of the sample mean" (with R. R. Rao), *Ann. Math. Statist. 31* (1960), 1015-1027.
19) "On the number of distinct values in a large sample from an infinite discrete distribution," *Proc. Nat. Inst. Sciences, India, 26,* A (Supp. 11), (1960), 67-75.
20) "A representation of the joint distribution of *n* dichotomous items," *Studies in Item Analysis and Prediction,* H. Solomon, ed., Stanford University Press, (1961), 158-168.
21) "On classification based on responses to *n* dichotomous items," *Studies in Item Analysis and Prediction,* H. Solomon, ed., Stanford University Press, (1961), 169-176.
22) "Classification into two multivariate normal distributions with unequal covariances" (with T. W. Anderson), *Ann. Math. Statist. 33* (1962), 420-431.
23) "On Fisher's bound for asymptotic variances," *Ann. Math. Statist. 35* (1964), 1545-1552.
24) "An optimal property of the likelihood ratio statistic," *Proc. Fifth Berk. Symp. Math. Statist. Prob., 1,* (1965), 13-26.
25) "A note on quantiles in large samples," *Ann. Math. Statist. 37* (1966), 577-580.
26) "Rates of convergence of estimates and test statistics," *Ann. Math. Statist. 38* (1967), 303-324. (This was a Special Invited Address to the Institute of Mathematical Statistics.)
27) "Substitution in conditional expectation" (with P. J. Bickel), *Ann. Math. Statist. 39* (1968), 377-378.

(28) "On conditional test levels in large samples" (with P. J. Bickel), *University of North Carolina Monograph Series in Probability and Statistics*, No. 3 (1970), 25-34.

(29) "Some asymptotic properties of likelihood ratios on general sample spaces" (with M. Raghavachari), *Proc. Sixth Berk Symp. Math. Statist. Prob., 1* (1970), 129-152.

(30) *Some Limit Theorems in Statistics*. NSF-CBMS Monograph, No. 4 (SIAM, 1971).

(31) "Examples of inconsistency of the likelihood ratio statistic," *Sankhya 34* (1972), 81-84.

(32) "A note on UMV estimates and ancillary statistics." *Contributions to Statistics* (Hajek Memorial Volume), Academia (Prague), 1979, 19-24.

(33) "On large deviations of the sample mean in general vector spaces," (with S. L. Zabell), *Ann. Probability, 7* (1979), 587-621.

(34) "Large deviations, tests, and estimates" (with J. C. Gupta and S. L. Zabell). *Asymptotic Theory of Statistical Tests and Estimation* (Hoeffding Volume), 1979, 33-67. Academic Press.

(35) "Hodges Superefficiency," 1980. *Encyclopedia of Statistical Sciences,* Vol. 3 (F-H). John Wiley.

(36) "On large deviations of maximum likelihood and related estimates." Tech. Report No. 121, Department of Statistics, University of Chicago, 1980.

(37) "A note on the effective variance of a randomly stopped mean." *Statistics and Probability: Essays in Honor of C. R. Rao,* 1982, 39-43, North-Holland Publishing Co.

(38) "Some further properties of the LR statistic in general sample spaces" (with T. K. Chandra and D. Lambert), 1982. *Proceedings of the Golden Jubilee Conference,* 1-19, Indian Statistical Institute, Calcutta.

(39) "Large deviations of the maximum likelihood estimate in the Markov chain case." *Recent Advances in Statistics* 1983, 273-286. Academic Press.

(40) "Distributional optimality and second-order efficiency of test statistics" (with J. C. Gupta), 1986. In *Adaptive Statistical Procedures and Related Topics,* Proceedings of a symposium in honor of H. Robbins, IMS Lecture Notes Monograph Series, Vol. 8, 315-331.

IMS Lecture Notes—Monograph Series
NSF-CBMS Regional Conference Series in Probability & Statistics
Order online at: http://www.imstat.org/publications/imslect/index.phtml

Vol	Title	Authors or Editors	Member	General
	IMS LECTURE NOTES-MONOGRAPH SERIES			
1	*Essays on the Prediction Process*	FB Knight	$9	$15
2	*Survival Analysis*	J Crowley & RA Johnson	$15	$25
3	*Empirical Processes*	P Gaenssler	$12	$20
4	*Zonal Polynomials*	A Takemura	$9	$15
5	*Inequalities in Statistics & Probability*	YL Tong	$15	$25
6	*The Likelihood Principle*	JO Berger & RL Wolpert	$15	$25
7	*Approximate Computation of Expectations*	C Stein	$12	$20
8	*Adaptive Statistical Procedures & Related Topics*	J Van Ryzin	$24	$40
9	*Fundamentals of Statistical Exponential Families*	LD Brown	$15	$25
10	*Differential Geometry in Statistical Inference*	SI Amari, OE Barndorff-Nielsen, RE Kass, SL Lauritzen & CR Rao	$15	$25
11	*Group Representations in Probability & Statistics*	PW Diaconis	$18	$30
12	*An Introduction to Continuity, Extrema & Related Topics for General Gaussian Processes*	RJ Adler	$15	$25
13	*Small Sample Asymptotics*	CA Field & E Ronchetti	$15	$25
14	*Invariant Measures on Groups & Their Use in Statistics*	RA Wijsman	$18	$30
15	*Analytic Statistical Models*	IM Skovgaard	$15	$25
16	*Topics in Statistical Dependence*	HW Block, AR Sampson & TH Savits	$27	$45
17	*Current Issues in Statistical Inference: Essays in Honor of D. Basu*	M Ghosh & PK Pathak	$15	$25
18	*Selected Proceedings of the Sheffield Symposium on Applied Probability*	IV Basawa & RL Taylor	$9	$15
19	*Stochastic Orders & Decision Under Risk*	K Mosler & M Scarsini	$18	$30
20	*Spatial Statistics & Imaging*	A Possolo	$21	$35
21	*Weighted Empiricals & Linear Models*	HL Koul	$18	$30
22	*Stochastic Inequalities*	M Shaked & YL Tong	$24	$40
23	*Change-point Problems*	HG Mueller & D Siegmund	$26	$45
24	*Multivariate Analysis & Its Applications*	TW Anderson, KT Fang & I Olkin	$26	$45
25	*Adaptive Designs*	N Flournoy & WF Rosenberger	$24	$40
26	*Stochastic Differential Equations in Infinite Dimensional Spaces*	G Kallianpur & J Xiong	$24	$40
27	*Analysis of Censored Data*	HL Koul & JV Deshpande	$18	$30
28	*Distributions with Fixed Marginals & Related Topics*	L Ruschendorf, B Schweizer & MD Taylor	$31	$52
29	*Bayesian Robustness*	JO Berger, B Betro, E Moreno, LR Pericchi, F Ruggeri, G Salinetti & L Wasserman	$29	$49
30	*Statistics, Probability & Game Theory: Papers in Honor of D. Blackwell*	TS Ferguson, LS Shapley & JB MacQueen	$32	$54
31	*L₁-Statistical Procedures & Related Topics*	Y Dodge	$41	$69
32	*Selected Proceedings of the Symposium on Estimating Functions*	IV Basawa, VP Godambe & RL Taylor	$42	$69
33	*Statistics in Molecular Biology & Genetics*	F. Seillier-Moiseiwitsch	$36	$45
34	*New Developments & Applications in Experimental Design*	N Flournoy, WF Rosenberger & WK Wong	$21	$35
35	*Game Theory, Optimal Stopping, Probability & Statistics: Papers in Honor of Thomas S. Ferguson*	FT Bruss & L Le Cam	$39	$66
36	*State of the Art in Probability & Statistics: Festschrift for Willem R. van Zwet*	M de Gunst, C Klaassen & A van der Vaart	$79	$120
37	*Selected Proceedings of the Symposium on Inference for Stochastic Processes*	IV Basawa, CC Heyde & RL Taylor	$29	$49
38	*Model Selection*	P Lahiri	$24	$40
39	*R.R. Bahadur's Lectures on the Theory of Estimation*	SM Stigler, WH Wong, D Xu	$15	$25
	NSF-CBMS REGIONAL CONFERENCE SERIES			
1	*Group Invariance Applications in Statistics*	ML Eaton	$15	$25
2	*Empirical Processes: Theory & Applications*	D Pollard	$12	$20
3	*Stochastic Curve Estimation*	M Rosenblatt	$15	$25
4	*Higher Order Asymptotics*	JK Ghosh	$15	$25
5	*Mixture Models: Theory, Geometry & Applications*	BG Lindsay	$15	$25
6	*Statistical Inference from Genetic Data on Pedigrees*	E Thompson	$18	$30